TITLE: EXPLORING THE GEOLOGY OF
SHELF SEAS

AUTHOR: McQuillin, R., Ardus.,
D. A., 1977

BORROWER'S NAME	DATE RETURNED

Exploring the Geology
of Shelf Seas

Graham & Trotman Limited

Exploring the Geology of Shelf Seas

R McQuillin   D A Ardus

Graham & Trotman Limited

First published in 1977
Graham & Trotman Limited
Bond Street House
14 Clifford Street
London W1, United Kingdom

© R McQuillin and D A Ardus 1977

ISBN 0 86010 012 X

Contents

List of figures   vii
List of tables   xi
Preface   xiii

1	Principal Objectives of Offshore Exploration	1
2	Position-fixing at Sea	7
	2.1 Introduction	7
	2.2 Co-ordinate systems and map projections	8
	2.3 Radio position-fixing systems	10
	2.4 Satellite navigation and position-fixing	24
	2.5 Integrated satellite navigation systems	27
	2.6 Seabed acoustic position-fixing	28
3	Echo Sounding and Scanned Sonar Methods	31
	3.1 Introduction	31
	3.2 The echo sounder	32
	3.3 Side scan sonar	39
	3.4 Bathymetric charts	45
	3.5 Geological applications	46
	3.6 Engineering applications	48
4	Continuous Sub-bottom Seismic Reflection Profiling	51
	4.1 Introduction	51
	4.2 Basic principles	52
	4.3 Seismic profiling equipment	54
	4.4 Analysis of profile data	73
	4.5 Geological applications	78
	4.6 Engineering applications	80
5	Deep Reflection and Refraction Seismic Methods	85
	5.1 Introduction	85
	5.2 Basic principles of multi-channel reflection seismic surveying	86
	5.3 Reflection seismic equipment	88
	5.4 Processing of seismic reflection data	99
	5.5 Analysis of seismic sections	103
	5.6 Applications	108
	5.7 The seismic refraction method	108

Contents

6   Magnetic Methods   113
6.1 Introduction   113
6.2 Basic principles   114
6.3 Survey magnetometers   116
6.4 Magnetic data reduction   120
6.5 Applications of the magnetic method   122
6.6 Geological interpretation of magnetic data   123

7   Gravity Methods   131
7.1 Introduction   131
7.2 Basic principles   133
7.3 The sea bottom gravity meter   134
7.4 The shipboard gravity meter   139
7.5 The Eötvös correction   145
7.6 Gravity data reduction   146
7.7 Applications of gravity surveys   149
7.8 Geological interpretation of gravity data   150

8   Sampling, Drilling and Visual Observations   157
8.1 Sediment sampling   158
8.2 Sampling outcrops and near-surface rocks   180
8.3 Sampling bedrock through a cover of superficial
    sediments   191

9   Developments and Requirements   199
9.1 The state of the art   199
9.2 Geological mapping   200
9.3 Hydrocarbon exploration   211
9.4 Seabed engineering   216

List of suggested reading   219
List of sources of technical information and illustrations   221
Index   228

List of Figures

Fig. 1/1   General scheme of offshore exploration operations.   3

Fig. 2/1   Range-range radio position-fixing pattern in an estuary.   12

Fig. 2/2   Hyperbolic radio positioning pattern derived from the same shore station layout as used to derive the range-range pattern in figure 2/1.   13

Fig. 2/3   Hyperbolic radio positioning pattern modified from that used in figure 2/2.   14

Fig. 2/4   The Decca Navigator System—English Chain (5B). Full daylight coverage and accuracy.   20

Fig. 2/5   Station disposition and coverage diagram for a Motorola Mini-Ranger III system.   22

Fig. 2/6   Orbit diagram for the Transit navigation satellites.   25

Fig. 2/7   Satellite pass giving six doppler counts.   26

Fig. 2/8   A 4-element doppler sonar.   27

Fig. 3/1   Echo sounder trace.   33

Fig. 3/2   Block diagram showing the main components of an echo sounder.   34

Fig. 3/3   Echo sounder recording system on electro-sensitive paper.   37

Fig. 3/4   Artist's impression of a side scanning sonar in operation.   38

Fig. 3/5   Side scan sonar record.   40

Fig. 3/6   Sonar slant range diagram.   41

Fig. 3/7   Slant range correction graph for side scan sonar.   41

Fig. 3/8   Block diagram showing the main components of a side scan sonar.   44

Fig. 3/9   Klein Hydroscan record showing exposed folding and faulting structures.   47

Fig. 3/10   Klein Hydroscan record showing a variety of patterns associated with different types of seabed sediments.   48

Fig. 3/11   Klein hydroscan record of the American schooner Hamilton.   49

Fig. 3/12   UDI AS Series side scan sonar record.   50

Fig. 4/1   Echo sounder record showing penetration to a sub-bottom rock layer.   52

Fig. 4/2   Block diagram showing the main components of a continuous sub-bottom seismic profiling system.   55

Fig. 4/3   Graphs of the relationship between seismic resolution and pulse-length and the relationship between resolution and source frequency for a 1-cycle pulse.   56

Fig. 4/4   Attenuation of seismic waves for an absorption of 0.5dB per wavelength.   57

Fig. 4/5   An Edo Western 515 pinger system.   58

Fig. 4/6   An Edo Western 515 pinger record.   59

Fig. 4/7   Pressure pulse produced by Huntec ED10 boomer.   60

Fig. 4/8   A Huntec ED10 boomer transducer mounted in a catamaran.   61

Fig. 4/9   An ED10 boomer record obtained in a North American Great Lake.   63

Fig. 4/10   Sparker acoustic sources; EGG 3-tip array and an IGS 200-tip array.   65

Fig. 4/11   A sparker profiling system manufactured by EGG.   66

Fig. 4/12   A sparker record from the Firth of Forth.   67

Fig. 4/13   A Multidyne hydrophone.   69

Fig. 4/14   An EPC Model 4100 graphic recorder with seabed sub-bottom seismic recording.   70

Fig. 4/15   A deep tow sparker record.   72

Fig. 4/16   Two intersecting sparker profiles.   74

Fig. 4/17   Secondary seismic events due to multiple and single point reflections.   75

Fig. 4/18   Plot of main seismic reflectors in Line 5 (figure 4/16) using equal horizontal and vertical scales and correcting for velocity variation with depth.   76

Fig. 4/19   Seismic profile showing site where bedrock possibly crops out at seabed.   79

Fig. 4/20   Seismic profiles across a range of seabed conditions in the vicinity of the Orkney Islands.   81

Fig. 4/21   Huntec deep tow boomer profile from offshore eastern Canada.   82

Fig. 4/22   NSRF deep tow sparker profile from offshore eastern Canada.   82

Fig. 5/1   Schematic diagram showing the use of multi-channel hydrophone streamers to acquire data which can be common depth point (CDP) stacked.   87

Fig. 5/2   Schematic diagram showing the principle of operation of four different non-explosive seismic sources.   89

Fig. 5/3   Signatures of a range of seismic sources.   93

Fig. 5/4   Schematic block diagram of the principal components of a Texas Instruments DSF V seismic recording system.   94

Fig. 5/5   Oscillograph record from a 24-channel streamer.   96

Fig. 5/6   Digital sampling of a seismic signal.   97

Fig. 5/7   Texas Instruments DSF V seismic recording system.   98

Fig. 5/8   LRS Manpaq seismic recording system fitted in Landrover.   98

Fig. 5/9   Part of a seismic section off N E Scotland.   100

Fig. 5/10   Velocity analysis showing coherence of stack at shot-point 830 in figure 5/9.   102

Fig. 5/11   Interpretation of the seismic section shown in figure 5/9.   105

Fig. 5/12   Interpreted profile from a survey of the Moray Firth.   105

Fig. 5/13   Isochron map of the base Jurassic in the Moray Firth.   106

Fig. 5/14   Refraction of seismic waves according to Snell's Law.   109

Fig. 5/15   Refraction of seismic waves in a three-layer earth.   110

Fig. 5/16   Layout of a seismic refraction experiment at sea.   111

Fig. 5/17   Results of a seismic refraction experiment in the Irish Sea.   112

Fig. 6/1   Variation in inclination of the earth's magnetic field.   115

Fig. 6/2   IGRF values of total intensity, F, in gamma for the epoch 1970.0.   117

Fig. 6/3   Schematic diagram of a marine survey magnetometer.   118

Fig. 6/4   A Barringer marine magnetometer.   119

Fig. 6/5   Aeromagnetic anomaly map of the Tremadoc Bay area of the southern Irish Sea. Contours in gamma.   121

Fig. 6/6   Top: Marine magnetic profile across a small seamount. Bottom: interpretation of a sparker profile over the same feature.   122

Fig. 6/7   Magnetic anomalies and interpretation offshore north-west Scotland.   127

Fig. 6/8   Computed magnetic anomalies over simple structures.   128

Fig. 6/9   Computed magnetic field across a horizontal prism.   129

Fig. 7/1   LaCoste and Romberg underwater gravity meter.   135

Fig. 7/2   Zero-length spring stress-strain graph.   136

Fig. 7/3   Schematic diagram of the LaCoste and Romberg gravity meter.   137

Fig. 7/4   LaCoste and Romberg air-sea gravity meter measuring system.   140

Fig. 7/5   LaCoste and Romberg air-sea gravity meter.   142

Fig. 7/6   Schematic diagram of the Askania sea gravity meter Gss3.   142

Fig. 7/7   Schematic diagram of the Bell marine gravity meter.   143

Fig. 7/8   Schematic diagram of a vibrating string accelerometer.   144

Fig. 7/9   Gravity data reduction flow diagram.   147

Fig. 7/10   Calculated gravity profiles over simple model structures.   152

Fig. 7/11   Calculated gravity profile across a model step structure.   153

Fig. 7/12   Bouguer gravity map of the Moray Firth. Contours in milligal.   154

Fig. 7/13   Computed (dots) and measured gravity profiles across margin of Moray Firth sedimentary basin.   155

Fig. 7/14   Comparison of observed and computed anomaly profiles across the Moray Firth for a computer-fitted model.   156

Fig. 8/1   Shipek Grab fitted with prototype sea bed camera and flash units undergoing initial sea trials.   160

Fig. 8/2   Schematic diagram of Shipek Grab.   160

Fig. 8/3   Shell sand waves traversing a gravel and pebble lag pavement   161

Fig. 8/4   Box corer. A stylised illustration showing the principle of operation.   162

Fig. 8/5   An IGS gravity corer system showing the half ton chassis unit.   164

Fig. 8/6   An IGS 20ft barrel vibrocorer.   168

Fig. 8/7   An IGS piston vibrocorer.   169

Fig. 8/8   Mode of operation of a piston vibrocorer.   172

Fig. 8/9   Disturbance due to the sampling process in cores obtained with the IGS vibrocorer.   174

Fig. 8/10   An outcrop (background) in an area of boulders with sediment occupying intervening areas.   183

Fig. 8/11   Schematic diagram of Pisces submersible.   184

Fig. 8/12   T.V. picture of sampling operation in Pisces submersible.   184

Fig. 8/13   The IGS submersible Consub showing the payload of small extendable rock drill, stereo camera system and T.V.   187

Fig. 8/14   Consub unmanned submersible showing payload.   188

Fig. 8/15   IGS 1m rock drill.   192

Fig. 8/16   Bedford Institute of Oceanography 20ft rock drill.   194

Fig. 8/17   The Maricor drill.   195

Fig. 8/18   *M.V. Whitethorn* drilling ship showing overside drilling platform.   197

Fig. 8/19   *M.V. Wimpey Sealab* drilling ship.   198

Fig. 9/1   Institute of Oceanographic Sciences 'Geological Long Range Inclined Asdic'–GLORIA.   201

Fig. 9/2   IGS geochemical eel.   203

Fig. 9/3   Comparison of processed and unprocessed sparker sections from the northern North Sea.   208

Fig. 9/4   Brent oil field.   212

Fig. 9/5   Comparison of conventional and bright spot processing.   214

Fig. 9/6   3-D seismic survey.   217

List of tables

Table 2/1   English Chain 5B basic data   19

Table 6/1   Magnetic properties of rocks   125

Table 7/1   Common rock densities   151

Preface

We wish to acknowledge our indebtedness to the Director of the Institute of Geological Sciences for approving the preparation of this book as well as giving his permission to publish a number of IGS photographs and to draw upon open-file records for some of our illustrations. The assistance afforded us by exploration and manufacturing companies in supplying information and illustrative material is also gratefully acknowledged. The current addresses of companies referred to in the text are supplied in appendix I.

A number of colleagues have given valuable aid by reading and commenting on parts of our manuscript, in particular D K Smythe, M Bacon and A S Mould. We are particularly indebted to Linda Nisbet, for secretarial assistance, and Angela McQuillin, for preparing the illustrations.

In our attempt to provide an account of so wide a subject as the methods currently being used in the exploration of continental shelf geology, we have become very aware of its incompleteness and of our omissions, but our aim has been to concentrate on topics not adequately covered in other text books, and in particular to present a practical rather than a theoretical approach to our subject.

R McQuillin
D A Ardus
Edinburgh
August 1976.

1 Principal Objectives of Offshore Exploration

As, with continuing exploitation, the reserves of non-renewable natural resources of the earth's land areas diminish, the need increases to win these same resources from beneath the seabed wherever they occur and can be economically exploited. Oil and gas, sand and gravel, coal, gold, iron ore, diamonds, manganese and nickel are some of the minerals and materials extracted from beneath the sea today. Most are extracted from beneath the relatively shallow seas adjacent to continental land masses. The geology of these shelf areas is of a continental (as opposed to an oceanic) type, and economic exploitation is largely within the control of the individual national governments of adjacent coastal states. In areas such as the North Sea international agreements have been reached defining the position of a median line which bounds the offshore territories of signatory nations, thus the way is prepared for commercial exploration and exploitation of sub-sea geological resources under a system of governmental control similar to that exercised in land areas. In other areas, negotiations between governments are being actively pursued.

Offshore exploration, exploitation of resources, and the engineering and construction of large seabed structures together form a major new and rapidly growing field of industrial development. A knowledge of the geology of the seabed and underlying strata is basic to this development and a knowledge of the methods used to acquire this knowledge is important to all who are either professionally involved, or training for such involvement. This book aims to explain the principles, applications and practice of methods used in offshore exploration. The nature of the problem is such that applied geophysics provides a major contribution to the exploration task and this bias is reflected in the content of the book. A number of texts already exist which adequately cover the theoretical basis of the main geophysical methods; some are suggested as further reading or for additional reference. Little, however, has been published giving an overall view of the nature of the practical problems to be encountered studying offshore geology. The marine environment is very different to that of the land; in some ways it makes exploration more difficult, in others more easy. The methods used are in many respects different.

In the chapters ahead, each method is treated individually and the applications to which it may be put are discussed. In practice, integration of a number of methods is usually necessary to solve a particular problem. For a complete geological interpretation, a combination of the results of all methods to be described must be amalgamated into an easily-understood geological synthesis. Written reports play an important role, but the main output from geological exploration is a series of maps. This series will include maps of bathymetry, of seabed sediment distribution, of thickness of superficial deposits, of bedrock geology, of geological structure, of isopachs to important horizons, of net thicknesses of important geological units etc. A general system for geological exploration is shown in figure 1/1. Many problems, particularly if linked to a single commercial objective, can be solved using only a limited part of this system. Everything depends on the aim of the work, the nature of the problem, and the finance available.

The oil industry has been, over the past decade, the major force behind development of exploration technology as well as providing most of the funds for the acquisition of data. Over this same period, research institutes and university departments have, with more meagre resources, played an important role, and it is on the results of research and exploration by such bodies that much published knowledge of offshore regional geology depends. Today, one of the more actively growing areas of the offshore survey scene is concerned with problems of site investigations for large-scale constructional and engineering work. Special problems of accuracy of location and precision of geological data arise in such work and methods are being developed to meet these needs.

Offshore exploration is costly. Operations at sea, even of a simple nature, involve use of large capital resources. Some examples can be quoted.

(i)   Inshore survey for pipeline route or coastal geological study: £300–£1000 per day.

(ii) Regional geophysical survey; bathymetry, gravity, magnetics, shallow seismics and sonar: £3000–£5000 per day.

(iii) Deep reflection seismic survey; usually quoted per line kilometre; £100–£200 including cost of processing. Daily rate about £5000–£10,000, ship and acquisition only.

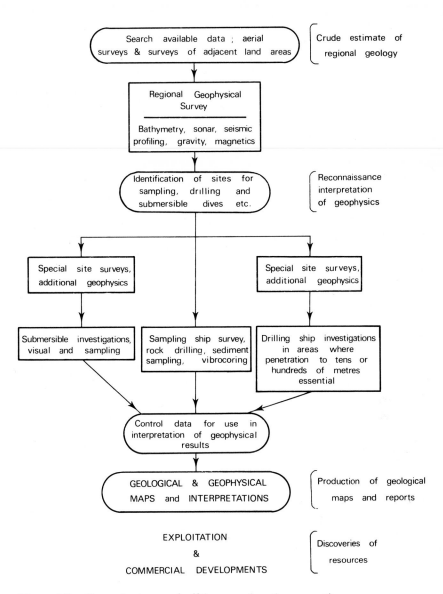

Figure 1/1   General scheme of offshore exploration operations.

(iv) Ship fitted for seabed sampling: £2000–£5000 per day.
(v) Ship fitted with dynamic positioning and drilling capacity to, say, 200–500m beneath sea bed: £10,000–£15,000 per day.
(vi) Semi-submersible drilling platform: £20,000–£30,000 per day.

The above figures are *very* general and can be considerably affected by a wide range of factors such as time of year and expected local weather conditions, depths of seas to be worked in, proximity to land and good service ports, factors concerned with conditions of contract, length of contract involved, required accuracies in position-fixing and other measurements to be made, and many others. Nevertheless it can be seen that when one considers daily costs, the offshore operation commences at a few hundred pounds per day and this can rise up to 100 times that cost. It is because of the high costs involved that planning is so important and that once a problem has been defined, the correct methods are chosen and properly applied.

Twenty years ago, in a book such as this, there would have been little to write about and few to write for. Now, the most difficult task facing the authors has been that of dealing adequately with their selected theme without indulging in too specialised a treatment of specific topics.

As a starting point, a chapter on position-fixing is included because a knowledge of positioning methods and their limitations is essential to an understanding of how the offshore survey is conducted.

Acoustic methods are of paramount importance; these are used to a limited extent in position-fixing and they are fundamental to studies of seabed topography using echo sounders and scanned sonar, in seismic reflection profiling of shallow sub-bottom layers, and in investigations of deep structure using digital seismic reflection as well as seismic refraction techniques.

The potential field methods—gravity and magnetics—play an important but less essential role. These methods are often used in reconnaissance surveys being less expensive than seismic work. More important is their role as an adjunct to seismic exploration. In this connection, gravity interpretation can solve problems of deep structure such as the tracing of major fault patterns and how basement structure controls the formation of structural elements at higher levels. Magnetic data can be used in detailed mapping of igneous and other magnetised formations, but again a more important role lies, perhaps, in the provision of regional information on the relationship between deep crustal structure and the genesis, history and structural form of sedimentary basins, the uppermost parts of which can be mapped in detail by the seismic methods. Over the past few years, the value of gravity and magnetic data has begun to assume an increasing importance in the regional interpretation of hydrocarbon provinces. Whereas, in early reconnaissance exploration of the North Sea, gravity surveys were not made by exploration companies whilst conducting seismic surveys, now, many speculative surveys being made around Britain are conducted to provide a package of data which includes seismic, gravity and magnetic results.

Geophysics, through various processes of interpretation, provides the geologist with a three-dimensional structural picture of the distribution of rocks and sediments at and beneath the seabed. To complete this picture he needs only to sample these rocks and sediments, study their composition, the fossils they contain, determine their age and chemistry, and ultimately make a stratigraphic identification (or estimate) and classify the samples as a particular type of rock or sediment. The geophysicist's structural interpretation is both incomplete and ambiguous until such geological control data can be incorporated into it. Unfortunately Man is not well equipped for field work on the seabed and the business of obtaining the desired samples is not simple. The scuba-diving field geologist is able to provide only a very limited contribution to the offshore exploration task. His endurance at the seabed is too short, his work is limited to shallow depths, and his mobility between sites is severely restricted. In many situations which are conducive to direct study, the diver still needs the fairly expensive backup of ship, support divers, and position-fixing equipment. His more mobile colleague, the geologist in the manned submersible, has at his disposal a technology which allows him to make a much more effective contribution to geological exploration, and much valuable work, including the collection of important rock samples, has been accomplished by geologists using small submarines. However, such an approach is costly, and routine sampling surveys over large areas of the seabed using such a method would be inordinately expensive. For this reason a wide range of tools has been developed which can be operated from an anchored or dynamically-positioned ship. The requirement is not simple and includes such tasks as that of removing a patch of seabed sediments a few centimetres thick from place of extraction to a sedimentology laboratory on land without excessive disturbance of the structure of this thin layer of unconsolidated material. Other tasks include that of obtaining sediment cores of a few metres length, of coring into bedrock, and in some circumstances that of coring through tens, or even hundreds, of metres of sediments and rocks. It is important that many of the sampling procedures should be linked to visual observations and a number of techniques have been developed for this purpose. Just as important is the need for close integration of all sampling and drilling work with previous geophysical surveys, through the selection of sampling and drilling sites on the basis of a preliminary geophysical interpretation, and the maintenance of adequate positioning accuracy during all phases of the exploration programme.

Advances in offshore exploration technology have been so rapid in recent years that this book can do no more than describe the state-of-the-art at the time of writing. In our final chapter, however, we look at some presently discernible trends towards development of new techniques viewed against our opinion of future demands for offshore geological data. In addition, some methods not described in the main text are briefly referred to such as the applications of radioactivity and geochemical exploration techniques; techniques which have so far had only limited application at sea when compared with their very wide application on shore.

Some topics have been omitted because in our view they are not sufficiently relevant to our principal theme of geological exploration; such topics include the techniques which have been developed for geophysical logging tests in boreholes, and the major technology of deep exploration and production drilling from large rigs and platforms. These subjects are briefly described in a companion volume: *Geology of the North-West European Continental Shelf, Volume 2*, by Pegrum, Rees and Naylor, within the more relevant context of exploration drilling for oil in previously identified prospective traps, and the use of geophysical logs to determine reservoir characteristics.

2 Position-fixing at Sea

2.1 Introduction

Any measurement or observation, whether geological or geophysical, whether at the sea surface or the seabed, whether made *in situ* or on a sample collected for laboratory study, if it is to be of value as part of a serious geological study, must be located (its position fixed) unambiguously with reference to a properly defined map co-ordinate system, and to a degree of accuracy commensurate with the objectives of the particular study or exploration programme being undertaken. In some cases a positional accuracy of 2km might be acceptable, in others an accuracy of 2m might be necessary. This chapter can only deal briefly with what is a very large subject, but position-fixing is such an important part of the business of work at sea, and good surveying is so crucial to the success of any exploration programme, that a basic understanding of navigation systems, their applications and limitations, is considered essential before embarking on any discussion of geophysical or geological exploration methodology.

The surveyor at sea does not have a topographic map showing the accurate positions of easily-distinguished landmarks to which he can refer local observations or sample sites, except in the particular case of surveys around coastal areas where visual fixing methods may be used. More usually his base-map is a drafted co-ordinate system and all positions are calculated relative to this system before plotting on the map. There are two aspects to the position-fixing requirement, one is the need to be able to return to a previously identified site, the other is a need to compile observations made at many different sites into maps or charts showing their correct spatial relationship. These needs must be separately considered when a project is being planned. With many position-fixing systems the accuracy with which a site may be relocated using identical deployment of positioning equipment is higher than the absolute positional accuracy of the site with reference to a world-wide co-ordinate system. It is always necessary to consider carefully the implications of accepting, at smaller initial cost, a low absolute accuracy simply because the system-related internal accuracy is sufficient to meet an immediate or short-term requirement. The long-term value of the data collected this way may be much reduced as a result of such a decision.

2.2 Co-ordinate systems and map projections

The first need of the offshore surveyor is a map on which to plot any surveyed position. But, to construct a map he must first devise some scheme whereby he may represent the curved surface of the earth on a flat piece of paper. A geographical map projection is just such a scheme. If the earth's surface were flat there would be no need to use a complex map projection system for representation of position, area, distance, direction etc. But the earth is not flat; its exact shape is not known, but it is generally considered to be best represented for mapping purposes as a sphere flattened at the poles, a shape which is described as an oblate ellipsoid of revolution, a spheroid. Most recent data on the shape of the earth have derived from observations of satellite orbits. Currently, according to internationally accepted values (the Reference Ellipsoid, 1967), the ellipticity of the earth is quoted as a flattening of 1/298.25, and the semi-major axis as 6,378,160m. The spheroid so defined is adequately accurate to allow construction of maps on a world-wide basis using a common projection system. It should be noted however that different mapping systems use small variations on these figures to describe the shape of the earth, the figures used reflecting the historical and geographical development of the particular system. In some areas, different spheroids have been adopted within single map systems as applied to construction of maps for different parts of the world. Two projection systems are most commonly used in offshore survey work: the Mercator Projection and the Transverse Mercator projection. A third system, the Lambert Conical projection forms the basis of some national co-ordinate systems, but has limited offshore application.

The Mercator projection system is traditionally that adopted for a use in construction of navigational charts except in polar regions. The reason is that using the Mercator projection, headings (compass directions of a ship's course), as measured on the chart are true, and a course at constant heading is a straight line. However, areas and shapes are badly distorted, and measurements of distance are subject to variable errors over the latitude range of any particular chart. The type of projection used in construction of these charts is termed conformal cylindrical; lines of longitude and latitude form parallels normal to each other, and the projection is that of the earth's surface onto a cylinder tangent to the earth at its equator. Such charts are often used from small-scale surveys such as harbour surveys and surveys of small areas adjacent to a coast. Again, the results of oceanographic cruises are sometimes plotted on 1:1,000,000 scale Mercator projection plotting sheets, but for large-scale survey and mapping projects Mercator projection charts are generally unsuitable.

The most common map projection used in offshore exploration surveying is the Transverse Mercator projection, a projection of the earth's surface onto a cylinder tangent at a meridian, that is a cylinder whose axis is normal to the earth's rotational axis, hence the name Transverse Mercator. The British National Grid system is a Transverse Mercator projection specially designed to allow preparation of maps of the land area of Britain to give a minimum distortion within a single co-ordinate system encompassing this entire area. Areas and distances are accurately represented, but offshore the system is only applicable over a limited E-W range. Beyond the eastern and western extremities of Britain distortions soon become unacceptable for survey purposes. For this reason, most offshore exploration work in the UK Continental Shelf, as well as in many other offshore areas around the world, is located on maps constructed using a Universal Transverse Mercator (UTM) projection. This sytem was introduced for NATO maps in the early 1950s. The world is split into numbered zones extending 6° in longitude, each zone extending from 80° S-80°N, numbering starting at the Greenwich antimeridian with zone 1 comprising 174°W-180°. Thus the zone between 6°W and the Greenwich meridian 0° is zone 30 and the zone between 0° and 6°E which includes much of the North Sea is zone 31. The origin of each zone is at the intersection of the central meridian with the equator and for work in the northern hemisphere this point is assigned the value 0 metres north (the northing), 500,000 metres east (the easting) so that all grid values in the zone are positive. For work in the southern hemisphere and equatorial regions the equator is given the value 10,000,000 metres north. The north-east tip of the British mainland, John o' Groats in Caithness, has geographical co-ordinates of approximately 03°W, 58°40′N; its approximate UTM co-ordinate location is given by 500,000E, 6,500,000N (UTM zone 30). The general acceptance in recent years of a world-wide map projection system linked to an internationally accepted spheroid and a series of agreed local datums (around Britain maps are based on the European datum) has done much to ease the task of the offshore surveyor whose life in earlier years was often plagued by the need to merge surveys conducted using different datums, map projections and co-ordinate systems. Furthermore, computer programs developed for reducing survey and geophysical data have now direct general application in different parts of the world. An added value of the general acceptance of UTM as a world-wide system is that it has eased the task of defining international boundaries and median lines between nationally designated areas of continental shelf and adjacent seas. In North Sea exploration, UTM is the accepted map projection and metric scales are used almost without exception. Most exploration work is plotted at a scale of 1:50,000, or 1:100,000; for broad reconnaissance surveys 1:250,000 scale is used. For detailed site surveys it may be necessary to use 1:10,000 or 1:5,000 scale maps.

Now, with map in hand, the task of the surveyor is that of locating the position of an exploration vessel, firstly with reference to certain known positions defined in the co-ordinate system and secondly to calculate the vessel's position in that co-ordinate system to a required degree of accuracy. Generally, two or more measurements of distance between the vessel and the reference points are the basis of the fix, but angles can be used as well.

We shall now briefly review the main methods and types of equipment available for position fixing.

2.3 Radio position-fixing systems

A wide range of radio position-fixing systems is available; these vary from world-wide systems using very low frequency signals which are propagated along the spherical wave-guide formed by the earth's surface and the ionosphere, to radar frequency and microwave systems which are effectively limited to line-of-sight operation. Accuracy depends on the stability and frequency of the system; great accuracy can only be obtained using stable high frequency systems. Range depends on the power and wavelength (inverse of frequency) of the transmission; long-range transmissions are only possible using very powerful, low frequency signals. Thus the ideal of long range and high accuracy is not obtainable and choice of a position-fixing system for a particular survey job will depend on the accuracy required, the availability or not of a suitable permanent system, the suitability of any available transportable system, range requirements and costs.

With one exception, Artemis, the radio positioning systems to be discussed below obtain fixes by determining distance values between survey points and shore or platform based reference stations.

Obviously, to fix a point, it is necessary to obtain at approximately the same time values of distance to more than one shore station. Two methods are used, the range-range method and the hyperbolic pattern method.

2.3.1. The range-range method

Sometimes called circular fixing, this method allows higher accuracy but less flexibility than the hyperbolic method. The most usual configuration is for an antenna on a ship to emit a signal which is transmitted to a shore station then re-transmitted to the ship, the round-trip being accurately timed and converted into distance by assumption of a propagation velocity for the radio waves (approximately 299,650km/s in air over sea water). Use of this method is usually restricted to a single survey vessel, mobilisation of the entire equipment system being dedicated to provision of a single position-fixing facility. For this reason, it is a specialised and costly method to employ. With some systems, however, the method can be used on a time-sharing basis so that more than one vessel equipped with master transmitter can use the same shore based stations. Generally, only very few vessels can be provided thus for concurrent operations. Even more flexibility is possible if a very high accuracy time standard is used both on ship and at the shore station. Recently, it has become possible to use Loran-C, see p. 16, in the range-range mode without any restriction on the number of survey participants. An atomic standard clock at the receiver is synchronised with similar clocks which control pulse transmissions from the shore stations. Single-way elapse times are measured from two shore stations to give a fix.

In figure 2/1 we see the method applied to a high precision, small area problem; the survey of an estuary. Shore stations are positioned on headlands A, B and C. It is assumed that the survey equipment can measure accurately (to say 1m) the distances from the ship's antenna to each of these stations. The requirement is that the survey vessel should be able to travel anywhere in the area of the estuary shown and obtain a good fix accurate to, say, 2m. Each fix will depend on plotting the intersection of two range circles; eg, at a point Z the range circle 6km from C and the range circle 6km from A intersect to give the fix position. It should be noted however that there is a second position of intersection Z′ which is also 6km from A and 6km from C. However, as this position is on land there is little likelihood of confusion. Such an ambiguity becomes most critical close to the straight line connecting AC (as Z and Z′ approach each other); fixes close to this line will have a poor resultant accuracy. Best fixes are obtained when range lines have an acute angle intersection close to 90° and preferably in the range 50°-90°. Thus, in the survey area illustrated, fixes east of C8 are best obtained using shore stations A and B, west of B8 using shore stations A and C, in the sector north of BDC using shore stations A and B, and in the sector C8 to B8 and south of BDC using shore stations B and C. No disposition of only two shore stations would give adequate positioning cover for the entire area. Although we have talked of plotting range circles in this discussion, more usually positions are plotted on pre-prepared charts with range circles already drawn, or the survey equipment is linked to a small computer which calculates and displays the position in terms of grid co-ordinates.

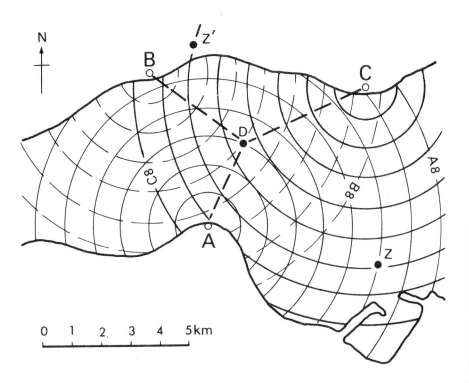

Figure 2/1   Range-range radio position-fixing pattern in an estuary.

2.3.2 The hyperbolic method

The hyperbolic method was developed to permit use of radio-navigation chains by unlimited numbers of vessels. It is employed in most long and intermediate range systems (world wide to, say, 200km). Short range, high accuracy systems more generally employ the range-range method. In the hyperbolic method, the receiving system on board ship is completely passive, hence the absence of any restriction on numbers of users of an established transmission chain. On board the survey vessel a receiver compares signals received from pairs of transmitting stations to derive the difference in distance of survey vessel from these two transmitters. A map shows a set of lines of position (LOPs) which are lines of equal difference in distance and describe hyperbola, see figure 2/2. This

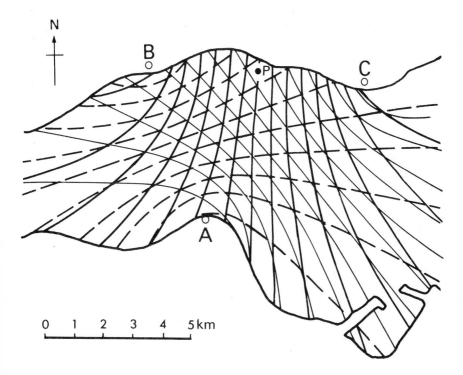

Figure 2/2   Hyperbolic radio positioning pattern derived from the same shore station layout as used to derive the range-range pattern in figure 2/1.

shows the pattern of hyperbolae derived from the same stations A, B and C as used to derive the pattern of range circles in figure 2/1. It can be seen that the area of good pattern cover (large angles between intersecting hyperbolae) is limited to the area within the triangle ABC. In this method, it is essential to have at least three shore stations, but these are capable of producing three sets of hyperbolic range lines. In our example, a better pattern could be produced by re-siting station A to a new position as shown in figure 2/3 thus extending good coverage to the harbour entrance. In planning survey work using either range-range or hyperbolic methods careful selection of transmitter sites can be a very important factor in provision of a good position-fixing facility.

A number of systems will now be described. It is not possible to mention every survey equipment on the market, and those referred

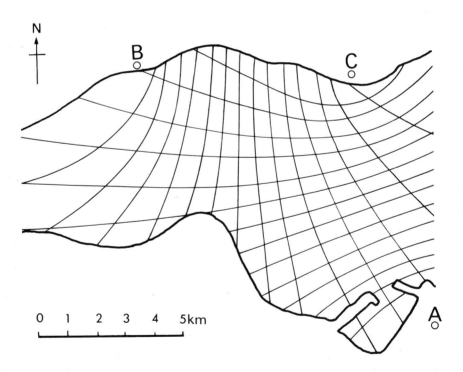

Figure 2/3 Hyperbolic radio positioning pattern modified from that used in figure 2/2 by re-siting station A to give good coverage to the harbour entrance in the south-east of the estuary.

to are selected because the authors have knowledge of their use in the offshore area around UK at the time of writing. The aim is principally that of covering the wide range of types of equipment available.

2.3.3 Long-range radio-navigation systems

The only truly long-range system giving global coverage is the Omega system which employs very low frequency transmission (10-14kHz) from eight stations in Norway, Liberia, Hawaii, North Dakota, La Réunion Island, Argentina, Australia and Japan. It is a hyperbolic system which depends for its long range capability on the fact that for such low frequencies the transmitted radio waves are 'trapped' between the earth's surface and the ionosphere. On board ship, an Omega receiver is tuned to a group of transmitting stations and by comparing the phase of signals from pairs of stations, hyperbolic line values are calculated and displayed. These values can be used then to obtain a fix on a pre-prepared lattice chart. For higher accuracy, corrections are applied for diurnal variations related to ionospheric changes; a computer is often used to apply such corrections and can be programmed to give a fix position in latitude/longitude values. With application of corrections, including correction information transmitted from a local land station, an accuracy of 1km is claimed. Without corrections an accuracy of 10km is possible, irrespective of time or place, on the earth's surface. One restriction to use of the method is that it depends on the instrument being set up at a known position (eg in a harbour) before commencement of a voyage and that the equipment operates continually until it can be re-set at another known position. It is usual therefore, for receiving equipment to incorporate certain 'fail-safe' features to ensure, as far as possible, the required continuous operation.

2.3.4 Medium-range radio-navigation and positioning systems

Omega is not accurate enough for the majority of continental shelf surveying tasks and its main use is in navigation not surveying. Medium-range systems include those used for both navigation and reconnaissance surveying: Loran-C, Decca Navigator etc; as well as systems designed for survey work: Decca Hi-fix, Pulse-8, Sea-fix, Toran and Loran etc. Like Omega, most of these systems are operated in the hyperbolic mode, though for specific projects, if the cost can be justified, a range-range adaption of the system can be used.

Loran-C is operated by the US Coast Guard with stations giving coverage which includes most of the northern Pacific, the east coast of the USA, North Atlantic, the Norwegian Sea and the Mediterranean Sea. It can be used up to distances in excess of 1500km from transmitting stations. It is generally used as a hyperbolic mode system, transmissions being as pulses not continuous waves. Each chain of transmitters consists of a master station and a number of secondary transmitters. All transmitters are controlled by atomic standard clocks, the timing of transmissions from the secondary station being adjusted for constant time difference to the master. A 100kHz carrier wave is used and fix accuracies of 100-200m are possible over wide areas and hundreds of kilometres away from transmitting stations. Modern receivers are relatively simple to use, and because the system works on a coded pulse principle, it is not so adversely affected by sky wave interference as are continuous wave systems. The system is not widely used in the North Sea because of poor reception over much of the area. It has been used in reconnaissance surveys around the Atlantic margins. A low-power variant called Pulse-8 (a Decca product) of the system has been introduced recently into the North Sea and is available as a transportable system. It is intended for use over ranges of less than 800km and aims to give an accuracy of 50m within a 500km range of any pair of transmitters, twenty-four hours per day.

Decca Navigator, often referred to as Decca Main Chain or Decca Mark 12, is operated by the Decca Navigator Company. Coverage includes north-west Europe and parts of Canada, the Middle East, Africa, India, Japan and Australia. The system is designed for multi-user application using the hyperbolic method; rental rates are relatively low and the receivers are easily operated. It is widely used for coastal navigation and by fishermen for re-location of fishing grounds, even by very small inshore fishing vessels. In areas of good line intersection and good signal reception accuracy can be sufficiently high for certain types of survey work. It is often used linked as a back-up to or component of integrated position fixing systems, for example as a link by computer to a satellite navigator. The system can be used up to about 500km from shore transmitter and gives an accuracy similar to Pulse-8, though unlike Pulse-8 it suffers from sky-wave effects which produce poor reception and variable errors during twilight and darkness periods. One feature of the system, not available with hyperbolic Omega or with Hi-fix, is that of line identification. In practice this means that the surveyor does not need to set up the equipment with reference to a known location before commencing a period of position-fixing. It is possible to switch on the receiver at sea given that position is known to say 15 km and almost immediately have indicated an unambiguous position within the accuracy limits of the system.

A Decca Navigator transmitting chain is made up of a master transmitter and three slave transmitters. The requirement is that the master and each slave should transmit a continuous wave of effectively identical frequency locked in phase. In practice, this is not done otherwise the receiver would be unable to resolve signals from the different transmitters. The effect is achieved as follows:

1) the master and slave stations transmit at harmonically-related frequencies;

2) transmissions from the master are monitored at each slave and a slave signal is then frequency controlled by the master signal and its phase locked to it;

3) the receiver uses multiplier stages to provide three comparison frequencies on which actual phase comparison measurements are made.

A typical chain is described in table 2/1, the data being supplied by Decca Survey Ltd for the two most common map projection systems used in Britain. It can be seen that the slave stations are given colour identifiers, and on lattice charts the hyperbolic lines of position (LOPs) are coloured red, green or purple to indicate which slave plus master combination of signals is used to generate the particular family of LOPs. These LOPs, or lanes, are also numbered on the charts and the receiver automatically registers lane number as well as giving a reading of LOP position to 1/100th of a lane. On the baseline (line joining master and slave) it is seen that 1/100th lane represents about 5 m. However, positional accuracies are always much less than 5 m.

Lane identification is achieved by generation of a set of coarse hyperbolic patterns for which the comparison frequency equals f. Lane identification signals are transmitted at fixed intervals during half second periods. Each pattern is identified separately in fixed rotation. During this interval the master transmits its normal 6f signal along with a 5f signal, the normal purple signal being shut off. Slave stations transmit a combination of 8f and 9f signals in turn. The lane identification sequence is transmitted once every minute with fifteen seconds between transmissions; f signals are obtained by subtraction (6f-5f and 9f-8f) in mixer circuits in the Decca Navigator receiver and lane identification information is displayed on a dial.

Table 2/1 English Chain 5B Basic Data

Ellipsoid: International (European Datum)*
  a = 6 378 388.0m
  e² = .006 722 670
* O.S.G.B. (1936) Datum.

CHAIN BASIC DATA	MASTER (6f) (Puckeridge)	RED (8f) (Shotisham)	GREEN (9f) (East Hoathley)	PURPLE (5f) (Wormleighton)
Latitude	51°54'37.65"N	52°33'12.12"N	50°54'38.67"N	52°11'53.20"N
Longitude	00°00'04.72"E	01°19'58.59"E	00°08'43.12"E	01°21'52.94"W
Rectangular X	3 942 996.47	3 885 392.81	4 029 922.78	3 916 644.03
Co-ordinates Y	90.22	90 406.95	10 220.53	−93 306.63
(Metres) Z	4 996 761.65	5 040 577.47	4 927 385.33	5 016 443.74
Baseline Length (Metres)		115 738.9	111 679.0	99 020.2
Transmitting Frequency (KHz)	85.0000	113.3333	127.5000	70.8333
Comparison Frequency (KHz)		340.000	255.000	425.000
Speed of Propagation (Kms/Sec)		299 250	299 250	299 250
Wavelength at Comp. Frequency (Metres)		880.1471	1 173.5294	704.1176
No. of Lanes on Baseline		263.000	190.330	281.261
First Lane		A0.000	A30.000	A50.000
Last Lane		2A23.000	2A40.330	1J61.260
Lane Width on Baseline (Metres)		440.0735	586.7647	352.0588

Ellipsoid: Airy
 a = 6 377 542.178m
 e² = .006 670 540

CHAIN BASIC DATA	MASTER (6f) (Puckeridge)	RED (8f) (Shotisham)	GREEN (9f) (East Hoathley)	PURPLE (5f) (Wormleighton)
Latitude	51°54'32.865"N	52°33'07.683"N	50°54'33.417"N	52°11'48.551"N
Longitude	00°00'05.471"E	01°19'59.989"E	00°08'43.974"E	01°21'52.787"W
Rectangular X	3 942 526.08	3 884 921.68	4 029 450.77	3 916 174.22
Co-ordinates Y	104.56	90 422.38	10 236.01	−93 292.54
(Metres) Z	4 996 189.04	5 040 007.02	4 926 810.49	5 015 871.74
Baseline Length (Metres)		115 740.8	111 679.3	99 020.4
Transmitting Frequency (KHz)	85.0000	113.3333	127.5000	70.8333
Comparison Frequency (KHz)		340.000	255.000	425.000
Speed of Propagation (Kms/Sec)		299 250	299 250	299 250
Wavelength at Comp. Frequency (Metres)		880.1471	1 173.5294	704.1176
No. of Lanes on Baseline		263.003	190.331	281.261
First Lane		A0.000	A30.000	A50.000
Last Lane		2A23.003	2A40.331	1J61.261
Lane Width on Baseline (Metres)		440.074	586.765	352.059

B

The Decca Chain described in table 2/1, the English Chain 5(B) has a master station north of London, the green slave near Brighton, the red slave near Norwich and the purple slave northwards of Oxford. This chain provides cover for the English Channel, the southern North Sea and some coastal waters east of England, see figure 2/4.

A range-range modification of Decca Navigator has been used to a limited extent, principally by the Hydrographic Department of the British Navy. The system is usually called 'Two-range Decca' and depends on siting the master transmitter on board the survey vessel.

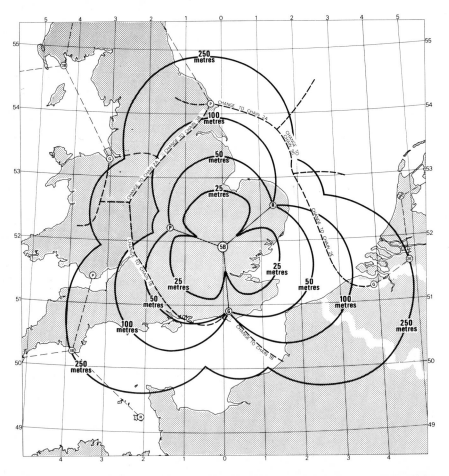

Figure 2/4   The Decca Navigator System—English Chain (5B). Full daylight coverage and accuracy. The full line contours enclose areas in which fix repeatability errors will not exceed the distances shown on 68% of occasions during full daylight conditions of time.

A greater accuracy can be obtained using the system in this mode particularly in surveys up to 400-500km offshore from a near-linear coastline.

The systems so far discussed, with the exception of Pulse-8, are marketed principally for navigation rather than surveying. The next system to be discussed, Decca Hi-fix, is a hyperbolic system operated at a frequency in the vicinity of 2MHz, designed and marketed almost exclusively for survey and similar applications where accuracies of approximately 20m are required. The range is less than that of Decca Navigator; 200km offshore being the typical limit for good reception. In areas such as the North Sea, a number of Hi-fix chains are operated by Decca on a continuous basis, but although it is a multi-user system, equipment and chain hire are still relatively expensive. With Hi-fix, both equipment and the mode of operation are very similar to that of Decca Navigator, except that Hi-fix chains do not have provision for lane identification. Thus it is necessary to commence any period of work with a setting-up procedure in which either the Hi-fix receiver is referenced against a known position such as a navigation buoy or drilling platform, or the survey vessel is navigated across baseline extensions to pass through known maxima and minima in LOP values. After referencing, it is essential that uninterrupted signals are recorded and care is taken to check that whole lane jumps do not occur on the receiver display during periods of poor reception. The existence of sky-wave interference and the lack of lane identification are the main drawbacks in this system which, nevertheless, is at present one of the most widely-used radio positioning systems around the UK. Currently, Decca are introducing a new version of Hi-fix, called Hi-fix 6, which offers greater flexibility as well as a type of lane-identification facility. Another Decca system, Sea-fix, is a transistoried version of Hi-fix which has had its main application as an easily transported system for use in shore-duration inshore surveys. Other commercial systems which function in a similar way to Hi-fix and which provide similar radio positioning accuracy include the French Toran and the American Raydist and Lorac systems.

2.3.5 Short-range radio positioning systems

For work demanding a higher accuracy than the 10m (approx.) of medium range systems or for work close inshore or in estuaries, short range, i.e. up to 35km, 'line of sight' systems are available. Portability is such that a complete system, including batteries, can be transported in a single estate car. Accuracies of between 1-3m can be achieved. Such systems are relatively inexpensive and easy to operate as single-user systems. Operational frequencies are in the same range as those used in radar. All, except Artemis, use the range-range method. Distances between vessel and shore stations are measured either by simple travel/time measurements or by phase comparison.

The Motorola Mini-Ranger III is an example of a system in which simple range measurements are made. In figure 2/5 a typical disposition of shore stations is shown. This shows the system as it was recently set up to give positioning cover over a group of offshore test ranges in the Firth of Forth. The Mini-Ranger operates in the C-band of radio frequencies at approximately 5500MHz, a wavelength of approximately 5cm. Aboard a survey vessel a transmitter/receiver is operated through an omni-directional aerial.

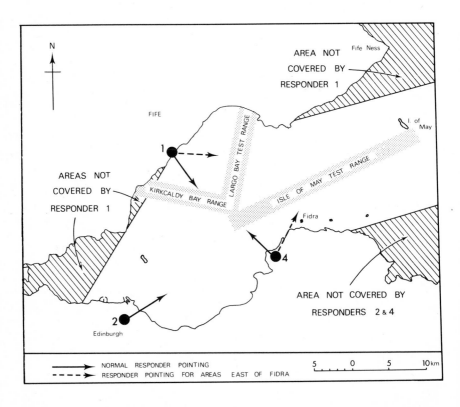

Figure 2/5   Station disposition and coverage diagram for a Motorola Mini-Ranger III system as used by the Institute of Geological Sciences for test range surveys in the Firth of Forth, Scotland.

Ranges are measured between vessel and usually two of the three available shore stations as follows:
1) The receiver/transmitter transmits a coded signal which activates a return signal from one of the shore stations;
2) the receiver computes and displays range in metres;
3) a second coded signal activates the second shore station;
4) the receiver computes and displays the second range.

In practice, these operations are repeated at rapid intervals and the values displayed are mean values of a group of valid replies. The system can be operated in conjunction with a desk computer and X-Y plotter so that a print-out of, say, UTM grid positions can be listed against time and a field plot of ship's position obtained. To conserve the batteries used as a power supply for the remote responders the main power-consuming circuitry of the shore stations is switched off when not in use. These circuits automatically switch on by interrogation from the shipboard transmitter and again automatically shut down after a period of no interrogation. The shore station antenna is not omni-directional; its 13dB sector has 75° azimuth and 15° elevation. In the Forth area, as shown in figure 2/5, most of the area of interest can be covered without any re-orientation of the antennae, nevertheless surveys cannot be conducted in the eastern part of the Isle of May range without re-orientation of stations 1 and 4 if the 75° sector limitation is assumed to hold. In practice, work in the area was possible without need to re-orientate responder 1.

A system very similar to the Motorola Mini-Ranger III is the Decca Trisponder system. The Decca system operates in the X-band at frequencies in the region of 9400MHz.

Systems which use phase comparison techniques include Cubic Autotape and Hydrodist. These systems attain a high accuracy of approximately 1m using S-band frequencies in the region of 3000MHz. The main operational difference between such a system and a travel-time system is that it utilises directional antennae and it is generally necessary to man the shore stations thus adding to operational costs. Greater ranges are obtainable and the systems are widely used to control the operation of pipe-laying barges, dredgers etc—very costly operations where high accuracy is essential. These systems have relatively few applications in geological and geophysical survey work.

In circumstances where high accuracies are essential but the survey area is far remote from a coastline, it is sometimes possible to utilise an offshore platform or an island as a single, fixed reference station. If such conditions apply, a range-bearing method must be used. One way of doing this for short-range work is to operate, for example, Hydrodist with one shore station and to place a theodolite at the shore station. A shore surveyor can observe bearings and radio values of these to the survey vessel. A fully-automated alternative is offered by an equipment called Artemis which is marketed by Decca. This operates in the 9200-9300MHz band and can be used to navigate up to 30km from the fixed station to give an overall accuracy of approximately 1.5m. Shore and ship antennae rotate like radar scanners until automatic alignment occurs, thereafter servo loops hold this alignment and the shore station relays bearing data to the ship. Range data is acquired in much the same way as with a Mini-Ranger or Trisponder system. The Artemis system can be linked to a small computer and XY plotter to give a continuous positional print-out.

2.4 Satellite navigation and position-fixing

Satellite navigation depends on reception of signals from a group of satellites operated by the US Government. These are called Transit satellites. The satellites and associated ground tracking stations have been in operation for the US Navy since 1964, and since 1967 data on the system has been available to allow commercial use. A wide range of receiving equipment is now marketed by a number of companies. At present, six Transit satellites are operational. These occupy circular polar orbits about 1075km high circling the globe every 107 minutes, see figure 2/6. As the earth rotates within this birdcage of satellite orbits, transmissions from a satellite pass are received on average at approximately 60-90 minute intervals. Passes are received least frequently at the equator, most frequently near the poles. Each satellite transmits signals at two carrier wave frequencies (150 and 400MHz) accurate to one part in 10^{11}. Each new transmission commences precisely on every even minute thus giving a precise time signal to the satellite receiver. The modulated signal provides data on the satellite's orbit to give its calculated position at the time of transmission. The orbital data are provided by a world net of tracking stations and orbit data are updated every twelve hours by transmission from a tracking station to each of the satellites. The survey receiver having recorded data on the location of the satellites at the commencement of a two-minute transmission, it remains to establish the position of the receiver relative to the satellite. This is achieved by making measurements of the doppler shift on received signals. If the receiver is stationary, say on a platform, the number of cycles received at the receiver in a defined period of time will differ from those that would be received if the satellite had been stationary, by a number which is a measure of the change of distance between satellite and receiver during the period of the count. This measurement gives a line of position (LOP) on the earth's surface fixed by the satellite position and doppler count value. During a satellite pass, five or more two-minute transmissions are received, as shown in figure 2/7, and during each transmission a number of doppler counts are made, giving a number of LOPs which theoretically should intersect at one of two points on the earth's surface symmetrically positioned to east and west of the satellite orbit. Usually

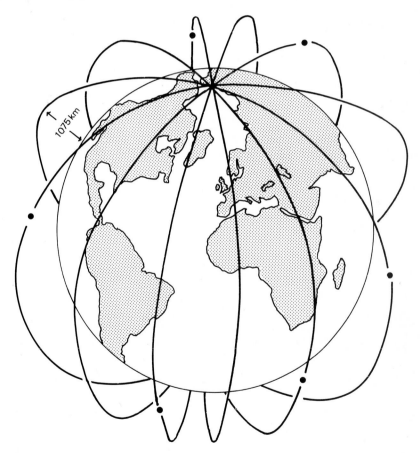

Figure 2/6   Orbit diagram for the Transit navigation satellites.

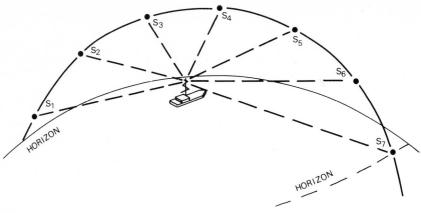

Figure 2/7   Satellite pass giving six doppler counts.

an approximate latitude and longitude value is fed to the computer before a period of operation to allow differentiation between these two possible positions. Results are processed by an iteration procedure and a solution for latitude and longitude is eventually derived. If the receiver is in motion, as on a survey ship, it is necessary to have a precise measure of the speed and heading of the ship as this affects the doppler count. If these parameters are accurately known, then a fix can be obtained to an accuracy of approximately 50m. Data on the ship's speed and heading, plus data on the ship's position between satellite fixes are obtained by integration of the satellite receiver, through a computer, with other navigational or position-fixing equipment. The main value of a satellite navigation system is that it provides reliable position-fixing data, with moderate accuracy, on a twenty-four hour basis anywhere in the world, and that once a ship is fitted with such a system, positional accuracy will depend only on the availability of a suitable secondary system to integrate with the satellite receiver. Geodetic position is controlled by the satellite system; the secondary system is required only to give relative motion data over periods of approximately one hour.

2.5 Integrated satellite navigation systems

For survey work in water depths up to approximately 350m, the type of system usually assembled has the following configuration:
Satellite navigation receiver,
Doppler sonar velocity sensing equipment,
Gyro compass,
Radio positioning system interface.
Such a system would normally operate using the doppler sonar to give an accurate measurement of the ship's speed over seabed (not through the seawater) and the gyro-compass to give an accurate measure of ship's head. All components are controlled by a computer, including the gyro-compass. The principle of the doppler sonar is illustrated in figure 2/8. A four-element sonar transducer array is lowered through the bottom of the ship to transmit four narrow-beam pulses; forward, aft, port and starboard. Comparison of received frequencies fore and aft gives a measure of the ship's forward speed, comparison of the port and starboard frequencies a measure of the ship's sidewards drift. The accuracy of such

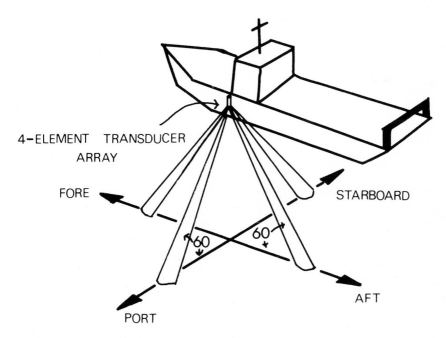

Figure 2/8   A 4-element doppler sonar.

measurements is increased by continuously monitoring the velocity of sound in seawater close to the sonar transducer array and applying this information to the calculation of speed. With such an integrated system in operation the computer can print-out ship's position at any pre-determined interval, say every minute, as well as prepare a complete log of the ship's position against time on digital tape.

At present, most sonar systems on the market are depth limited to approximately 350m. In deeper water, or in circumstances where fitting a sonar is impracticable, the satellite navigator can be integrated to a radio positioning system such as Loran-C or Decca Navigator. Alternatively, the sonar can be operated by reflections from within the body of seawater, in which case the recorded ship's speed is that through the water. The computer is then used to calculate an estimate of sea current velocity and direction to compensate for this effect. It is also possible to integrate satellite and inertial navigation systems but costs are very high and can be justified at present only for extended surveys in deep water areas.

It should be noted that values of latitude and longitude as derived from Transit satellite data are in terms of a satellite derived world geodetic system and that it is necessary to apply corrections to obtain positions in terms of a local land system such as the British National Grid system, UTM etc. In the middle of the North Sea, for example, a satellite position is separated from a UTM position by approximately 120m. Corrections are usually applied by computer before print-out.

2.6 Seabed acoustic position-fixing

Position-fixing with reference to fixed points on the seabed can be obtained by deploying a number of acoustic transponders on the sea floor in the immediate vicinity of a target area. This method is used, amongst other applications, for tracking submersibles during survey work, for accurate seabed surveys of platform and rig sites, for pipeline surveys and as a means of bringing rigs and platforms onto pre-determined site positions. The sytem works in range-range mode, therefore only two transponders are necessary to give a position. In practice, usually three or four transponders are used so that good range-circle intersections are obtainable over an entire survey area. Fixes can be obtained up to about 5-10km from the most distant transponder. In a typical system, all transponders are activated by a single interrogation pulse from the survey vessel, whether ship or submersible, and within a group each transponder responds at a different frequency. In a system called Autranav, marketed by Polytechnic Engineering, a computer controls an integrated satellite navigation and sonar transponder system. Transponders are deposited on the seabed in positions, known only approximately, to cover a particular investigation. Over a period of hours in the area, seabed geometry is determined by sonar measurements whilst the satellite navigator logs a succession of passes progressively obtaining a more accurate determination of the absolute position of the transponder group. Once the group has been calibrated, they continue to function for up to three years, and at commencement of a new survey it is immediately possible to detect any displacement of a transponder due, for instance, to the action of a fish-trawl. Accuracies claimed for such a system are of the order 5m for seabed location, 10-15m absolute. Such a system is ideal if accurate re-location on the seabed is of primary importance. The system has great value also if high accuracy surveying is required at distances beyond the range of accurate radio navigation systems.

In conclusion it may be said that choice of the best position-fixing system for a particular task is a complex one, but of paramount importance to the success of a project. Position-fixing technology is still undergoing rapid development; many of the systems described here have either arrived on the market during the past five years or have been subject to extensive improvement during this period.

3 Echo Sounding and Scanned Sonar Methods

3.1 Introduction

A survey of the topography of the sea floor is essential to almost every exploration programme and is usually conducted either in advance of, or at the same time as, other surveys. Results are usually presented as bathymetric charts. Around the British coasts, through the work of the Hydrographic Department of the Navy and of Port Authorities and Docks Boards, a full coverage of bathymetric charts is available to the public on a wide range of scales; indeed such charts are available covering most of the world's shallow seas usually through government agencies and in particular for those areas regularly used by the international shipping trade. However, these charts have been principally designed in the past for the use of navigators, and in exploration work more detailed or more up to date information is often required; furthermore, it may be of importance to make depth measurements concurrent with measurements of other parameters. Published navigational charts which contain bathymetric data are of most value during the planning stage of a programme when decisions must be made as to size and type of vessel to be used, layout of survey lines, anchoring capability of survey ship, depth capability of devices planned for operation on the sea floor, and many others.

As well as being needed to control other surveys and to provide the topographic base in engineering work, drilling, pipelaying etc, information on seabed topography can be a useful aid to interpretation of seabed geology, and also can give an indication of the distribution of different types of sea floor sediment. This aspect of the study of seabed topography will be considered later in this chapter.

Two methods are available for observing seabed topography. Echo sounders are used to make discrete measurements of depth below floating vessels, and from profiles of such measurements water depth charts can be constructed. This method gives a topographic profile along a section of the sea floor directly beneath the survey ship. Even if a network of such lines is surveyed, considerable interpretation is required if the echo sounder data are to be contoured correctly and a meaningful picture of seabed topography constructed by interpolation between traversed lines. As a rule the offshore surveyor is not able to make visual observations of the seabed to aid this process of interpolation. However, in recent years a tool has been designed, which is in principle similar to radar but uses instead of radio waves a pulsed beam of sound which is scanned through the underwater darkness to detect changes in seabed topography, variations in seabed materials, and the positions of wrecks or pipelines where these protrude above the sea bottom. This tool is the side-scan sonar, and it is used in a wide variety of situations including that of supplementing the data acquired by echo sounder surveys in the construction of bathymetric charts.

3.2 The echo sounder

Echo sounders do not provide direct measurements of water depth. A pulse of sound is emitted by the sounder and an echo from the seabed detected. The time interval between transmission of the pulse and detection of the echo is measured. This pulse of sound has travelled to the seabed and back and the time interval is called the two-way travel time.
Thus:

Depth = ½ vt,
where t = two-way travel time
and v = velocity of sound in water.

Echo sounders are calibrated for an estimated average velocity of sound in water and usually give a direct display of the profile of depth changes along a traverse line surveyed over the seabed. Figure 3/1 shows a typical echo sounder record. Very large differences in horizontal and vertical scale are usually a feature of such records and this leads to a gross vertical exaggeration of topography. In the example, figure 3/1, the feature in the middle of the record is where an igneous dyke crops out at the seabed. In this case the vertical exaggeration is approximately the order of magnitude of twenty.

The type of echo sounder to be selected for a particular project will

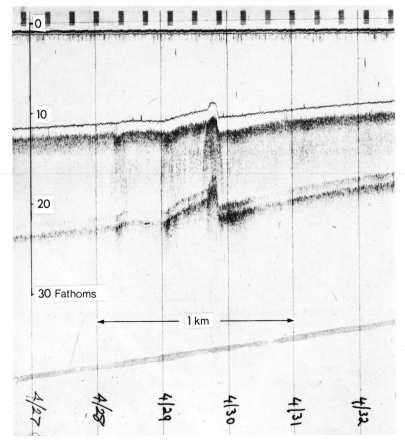

Figure 3/1   Echo-sounder trace.
(Photo: IGS.)

depend on a number of factors: precision and accuracy required, depth range anticipated, maximum depth of water in area to be surveyed, whether or not a good quality graphic record is required for geological or geotechnical interpretation, and, whether or not a digital output of depth is required to be logged by ship computer or data logger. Thus capital cost can range from a few hundred to a few thousand pounds per instrument depending on requirements.

The main components of the instrument's system are as follows: a transmission unit, transmitting and receiving transducers (sometimes separate, sometimes one transducer is used for both purposes) a receiving amplifier and signal processor, and a recorder or display unit, see figure 3/2.

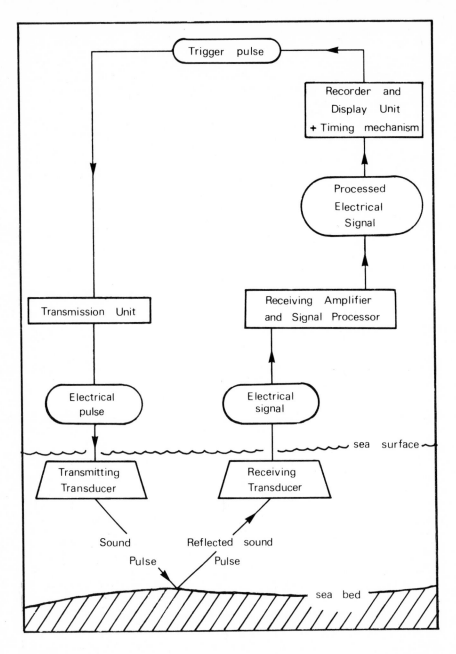

Figure 3/2 Block diagram showing the main components of an echo sounder.

The transmission unit consists of a signal generator and associated switching circuitry, generating a pulse of alternating current electrical energy which is fed to the transmitting transducer. The frequency of the electrical pulse is tuned to the natural frequency of the transmitting transducer for maximum efficiency in conversion of electrical energy to sound energy.

Transmitting and receiving transducers are usually identical, and in some instruments, through use of special switching circuitry, the same transducer is utilised for both functions. Two types of transducer are in use, magnetostriction and piezoelectric. Magnetostrictive transducers are energised by pulses of relatively low voltage alternating current. They are capable of handling more power than equivalent piezoelectric transducers and can be designed to transmit sound pulses through a ship's bottom plate, thus fitting-out and maintenance are relatively simple jobs and do not involve docking the survey vessel. Such transducers are not efficient convertors of electrical energy to sound and the consequent high power consumption can be a disadvantage of such systems for some applications. Furthermore, they cannot be designed to operate well at high frequencies. Piezoelectric transducers are very efficient in converting electrical energy to sound and operate well at high frequencies but cannot handle very high power outputs. They must in general, be mounted in direct contact with the sea, either recessed into the ship's hull, fitted into a towed fish, or mounted into an assembly which is fitted to the side of a survey vessel so that the transducers can be lowered beneath its waterline for survey work and hauled inboard for maintenance and to avoid damage in harbour.

In practical terms, the choice of transducer depends on a number of considerations including the power available for consumption and the power output required, permanency of the assembly and requirements related to output frequency. It is these variables which control the accuracy, resolving power and depth range of the instruments. High frequency devices can be designed to give higher accuracy and better resolution of small seabed features, whereas low frequency devices can be used to obtain sub-bottom as well as bottom reflections. Instruments in use today range in operating frequency from approximately 2kHz to a few hundred kHz.

The receiving amplifier converts the weak electrical signal received from the receiving transducer into electrical signals which will correctly activate the recording and display units. Various controls are incorporated into the circuits so that the quality of display can be optimised in much the same way as volume, tone and tuning controls are incorporated into a radio to control the quality of sound emitted by its loudspeakers. The nature of the processed signal will depend on the type of recorder used. Wet paper recorders require low voltage signals, whereas high voltages are required to mark the electro sensitive dry-recording papers more commonly in use today.

It is basic to the operation of all echo sounders in common with all other acoustic exploration systems, that recording and display units are precisely synchronised with the transmitting unit, and that the system is controlled by an accurate timing device. Therefore, it is essential that the time of transmission of a pulse from the transmission unit is controlled by the recording unit.

The picture seen on an echo sounder record is built up by sweeps of a stylus across the recording paper in a way similar to that by which a television picture is built up by sweeps of an electron beam across the face of a television tube. In the echo sounder, after each sweep of a stylus the paper moves a small increment so that the mark of the next sweep abuts with and to some extent merges with the previous one, and a succession of marks is built up to give a continuously marked profile of the sea surface to seabed interval. The recording system is illustrated in figure 3/3. A belt fitted with a number of styli transports these across the paper at a constant speed with a constant separation A-B. As each stylus passes point A a trigger pulse activates the transmission unit which in turn activates the transmitting transducer sending out a pulse of sound towards the seabed; and at the same time the paper is marked at point A. During the time interval taken for the stylus to progress from A to C the sound pulse is beamed to the seabed, echoes back, and is received at the receiving transducer. At C, the echo, having been converted into an electrical signal, marks the paper again. The sweep speed of the stylus is so calibrated that the distance A-C corresponds to the correct depth interval as marked on the echo sounder's measuring scale. When the stylus reaches B, another stylus reaches A, and by this time the paper has moved a small increment forward.

To register the relationship between ship's position and the echo sounder trace, at time intervals usually of a few minutes, the record is marked with a line such as lines X, Y and Z in figure 3/3. At each such instant of time the ship's position is recorded in a survey log and marked on a chart. Thus lines X, Y and Z on the echo sounder record correspond with known positions of the ship as it progresses along a traverse.

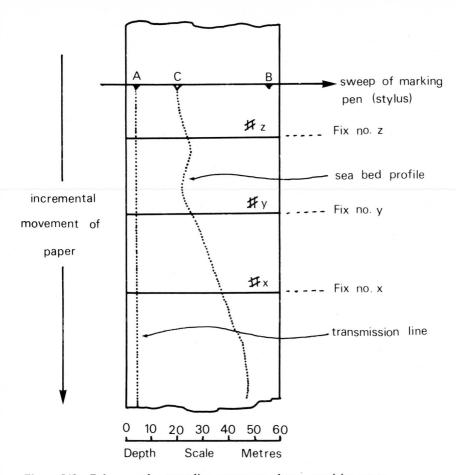

Figure 3/3   Echo sounder recording system on electro-sensitive paper.

The first stage in the production of bathymetric charts is then to transfer depth values measured for each fix position onto the survey map, the depth values being termed 'posted' values. Also at this stage, depth values intermediate between fixes are usually posted, in particular, to mark topographic highs and lows as seen on the echo trace. Once a grid of lines has been surveyed in an area it is possible, if required, to contour the data to produce a contoured bathymetric chart. However, it may first be necessary to apply corrections to the measured depth values to compensate for tidal effects, to adjust to a pre-defined datum, and to compensate for variation with depth of the velocity of sound in water. Such refinements will be discussed in section 3.4, but before doing so we will first consider the side scan sonar method.

Figure 3/4   Artist's impression of a side scanning sonar in operation. (Courtesy of Klein Associates.)

3.3 Side scan sonar

The side scan sonar method was developed in the late 1950s from experiments using echo sounders tilted away from the vertical. Such sounders were studied as a possible means of detecting shoals of fish, but results also showed the potential of the method for studying the geology of the seabed and the detection of wrecks as well as natural features of seabed topography adjacent to, but not directly beneath a ship's course. Modern equipment utilises specially designed transducers which emit a focused beam of sound having a narrow horizontal beam angle, usually less than 2°, and a wide vertical beam angle, usually greater than 20°; each pulse of sound being of very short duration, usually less than 1ms. To maximise the coverage obtained per survey line sailed, systems have been designed which are dual-channel, the transducers being mounted in a towed fish so that separate beams are scanned to each side of the ship. Thus a picture can be constructed of the seabed ranging from beneath the ship to up to a few hundred metres either side of the ship's course. The range of such a system is closely linked to the resolution obtainable and in this book we will concentrate on high resolution, relatively short-range systems (100m-1km per channel) as these are the tools commonly used in continental shelf exploration.

Typically, a high precision system would be towed some 20m above seabed and operated to survey to a range of 150m either side of the ship, and this configuration is illustrated in figure 3/4. In figure 3/5 an example of record is shown obtained over a variety of seabed morphological and geological conditions.

As with the echo sounder, the basic principle is that of detecting echoes of a transmitted pulse and presenting these on a facsimile record, termed a sonograph, in such a way that the time scan can be calibrated in terms of distance across the seabed. The first echo in any scan is the bottom echo, subsequent echoes being reflected from features ranging across the seabed to the outer limit of the scan. A number of points should be noted. The range scale shown on a sonograph is usually not the true range across the seabed but the slant range of the sound beam, A in figure 3/6, and that as with the echo sounder, distances indicated on a record depend on an assumption as to the velocity of sound in water, distance = $\frac{1}{2}vt$. Thus, if properly calibrated, the sonograph will show a correct value for B, depth of water beneath fish, and this depth profile is usually easily seen on a good quality sonograph. Echoes reflected across the scan, subsequent to the seabed echo, from points X

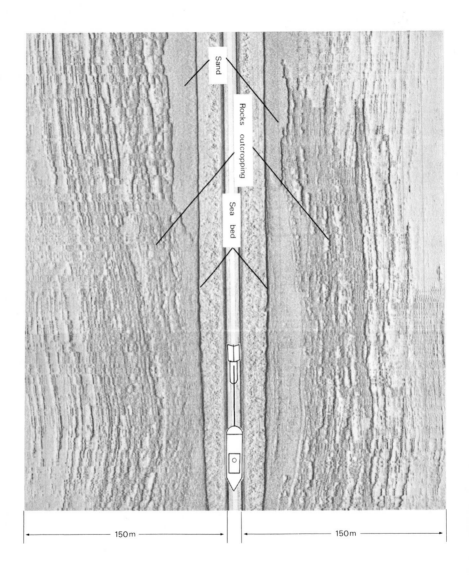

Figure 3/5   Side scan sonar record obtained offshore south-east Scotland with a Klein Associates hydroscan equipment.
(IGS record.)

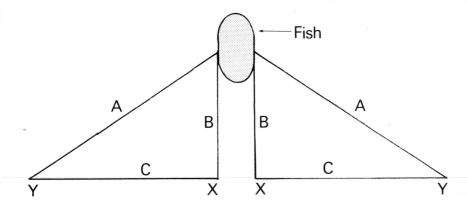

Figure 3/6 Sonar slant range diagram.

to Y in figure 3/6, are subject to slant range distortion: the actual distance scanned across the seabed is $C = \sqrt{A^2 - B^2}$. In our example in figure 3/5, in 20m water depth, $C = \sqrt{150^2 - 20^2} = 149m$, wheareas the scaled seabed record width is $150-20 = 130m$. Thus, if an object is detected by side scan and it is required to make a precise measurement of its size and position relative to a fixed position of the ship, corrections for slant range distortion should be applied. If B and A are to be kept near constant for a survey, a correction graph can be constructed relating apparent range to true range and one such graph is shown in figure 3/7.

It can be seen from this graph that most of the distortion is in the zone immediately beneath the towed fish, and in the range 50-150m

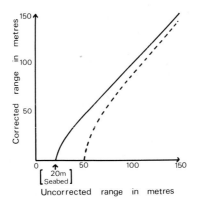

Figure 3/7 Slant range correction graph for side scan sonar towed at 20m (solid line) and 50m (broken line) above seabed.

distortion is negligible. However, if the fish were towed at 50m above the seabed, the dashed line curve in figure 3/7, the distortion would be more significant over a wider part of the scan, and uncorrected range values would be significantly different from corrected values over the full range of the scan. In practical terms, if it is possible to tow the sonar fish above seabed in the vicinity of 10-20 per cent of the full scale setting chosen for an operation, then the sonograph pictures will be free from excessive non-linear slant range distortion except in the region very close to the line of profile. Sonar fish housings are usually designed on the assumption that these conditions are approximately met, with the transducers set with vertical beam tilted down at 10° from horizontal.

In some respects, a more important type of distortion to which sonographs are subject is that of scale distortion. Distances measured along the sonograph are usually not equal to distances measured across the scan direction. In the example in figure 3/5, the ratio of scales along course to across course is approximately 1:3. In some cases it is possible to obtain near isometric presentation by varying the paper drive speed, but some equipments do not have this facility, and often the quality of record would be worsened if such a solution were attempted. If it is necessary to make mosaics of the seabed in the same way as in aerial photography, then records must either be photographed using a non-linear photographic technique, or side scan data must be recorded on analogue magnetic tape for replay through a display device which can be programmed to give isometric presentation at a selected mapping scale: both oscilloscope and fibre-optic recorder based systems have been developed for this.

Side scan sonar techniques being much more specialised than echo sounding, the range of equipment at present available to the operator is quite narrow. Most equipment aims to meet the same main requirements for work on the continental shelf where the widest market is with companies engaged in engineering studies in water depths up to about 200m. Perhaps the most important variable to be considered is the resolution required. For highest resolution, a high frequency sound source, possibly in the range 50-500kHz, and a very short pulse length, say 0.1msec, is required and such a source should be capable of a range resolution of 20-50cm and detection of small scale features of seabed morphology such as sand ripples of 10-20cm amplitude. However, maximum range of such a system is not likely to exceed 400m. If lower resolving power is acceptable, systems based on lower frequency sources are available which can be operated successfully over larger sweep ranges. Thus, if the object of a survey is to obtain complete sonograph coverage of an area, range limitation will be an important factor in the cost of the undertaking.

The configuration of main components of the instrument system is very similar to that of the echo sounder, though with dual channel systems each channel constitutes a single channel sub-system consisting of transmission unit, transmitting and receiving transducers, receiving amplifier and signal processor, see figure 3/8. Both channels then feed a common recording, display and time control unit.

The design of a sonar transmission unit aims to produce a high voltage short duration pulse to activate the transmitting transducer, and capacitor discharge techniques have been developed for this purpose. The pulse causes the transducer to oscillate at a high frequency at large power level but over only a few cycles.

Transmitting and receiving transducers are constructed as line arrays of piezoelectric elements (magnetostrictive elements can be used in lower frequency devices) and in a dual channel system four such arrays are located in the tow fish. The design and characteristics of tow cable and fish have a large effect on the quality and flexibility of operation of the complete system. In the cable, receiver leads need to be individually screened to eliminate crosstalk between channels and the cable needs to have a high breaking strain rating yet be reasonably flexible for deck handling. For deep tow systems a few hundred metres of cable need to be streamed when in operation. The tow fish must swim without yawing as this would distort the sonar pictures and the tow fish/cable assembly must be capable of depression at survey speed to a depth of within a few tens of metres of seabed and be towed at this depth along an approximately horizontal plane. These requirements have been met in different ways by different manufacturers, but all methods involve fairly difficult handling procedures, a situation often exacerbated by the fact that the sonar fish is an expensive item, costing upwards of £2000, to lose through collision with the seabed.

The function of receiving amplifier and signal processor in a side scan sonar is similar to that of the equivalent unit in the echo sounder (see section 3.2) but as we are now concerned with signals other than those due to first arrival echoes from the seabed, a more complex signal processing facility is required, and in particular time variable gain is necessary to ensure that signals due to echoes from the far range of a sonar scan are printed on the facsimile record with the same order of intensity as those due to echoes from targets close to the sonar fish. In the past few years, considerable advances have been made in signal

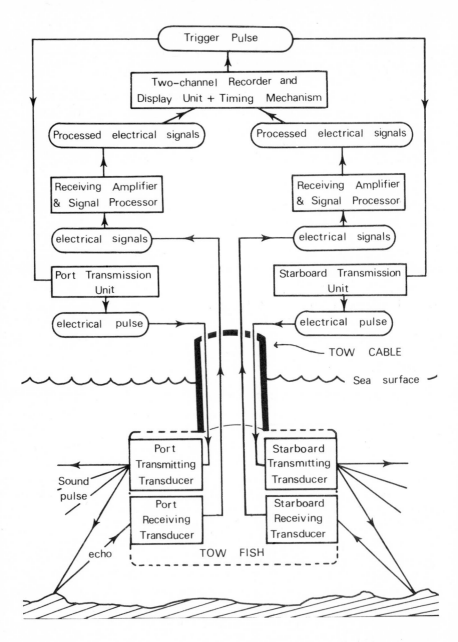

Figure 3/8   Block diagram showing the main components of a side scan sonar.

processing techniques. Early sonar instruments suffered from a number of defects which made it difficult to obtain consistency in record quality, both across scan and along profile with variation of water depth and seabed conditions. Such instruments demanded much attention and knob-twiddling for best results.

The recording and display unit, with associated timing mechanism is again similar in operation and function to that described in section 3.2 for echo sounders, except that being in most instruments a dual channel system, a means has to be devised for split-trace or dual-trace printing so that both channels are printed onto the same roll of recording paper. One way of meeting this requirement is to use a recorder fitted with two helical electrodes wound with opposite pitch on the same drum; the record shown in figure 3/5 was obtained from such a recorder. Sonar records are fix-marked in the same way as echo sounding traces, and the position of important targets can be transferred to a survey map applying corrections for slant range distortion, if these are significant.

3.4 Bathymetric charts

The term 'bathymetric chart' is in general use for a range of maps and charts showing measurements of topographic variation on the seabed. Like any other topographic map, it is important to note the datum to which measurements are referred. The datum selected is usually chosen to meet the needs of the potential user. If the map is to be used for navigation then a local chart datum is usually used which is approximately the level of mean low water springs. If the map is to be used for geological or engineering purposes, and it is possible to do so, measurements are usually referred to the local land survey datum—the Ordnance Datum (Newlyn) in Britain—by linking the hydrographic survey to a land survey benchmark. The Newlyn datum is based on a levelling mark in Newlyn Harbour which was established after a period of measurement of mean sea level there.

Topographic surveys at sea cannot be made to the high accuracies obtainable on land. Depth measurements from a ship are made with reference to a sea surface level which rises and falls in response to varying tidal and meteorological conditions which can in most circumstances be corrected for but not to a high degree of accuracy. In some survey areas, depending on the local tidal range, the methods used and survey requirements, corrections for this rise and fall are subject to errors and uncertainties which are almost as large as the predicted rise and fall values. In such circumstances, maps and charts are usually rendered uncorrected, the costs of obtaining adequate correction data being unwarranted. Thus although it is generally quite practicable to make accurate surveys of estuaries and coastal areas with reference to a geodetic levelling datum, it is seldom practicable to extend such surveys far from the coastline without resultant loss of accuracy, even though the precision with which water depth is measured may remain constant.

In offshore geological and geophysical exploration, depth measurements are usually made concurrently with other tests or measurements. It is basic to geological investigations that samples, boreholes and underwater photographs are reliably located both in terms of geographical position and depth. Thereby the effects of topography on sediment distribution can be determined as well as the structural relationships between widely spaced samples of solid rock. In geophysics, depth values are needed to an accuracy of approximately 1m for the application of corrections for the effects of a water layer in processing gravity and seismic data. In some circumstances lower accuracies can be tolerated.

Apart from the need for depth information to define location, echo sounder and side scan sonar surveys have great value as an aid to interpretation of seabed geology, a subject discussed in the following section.

3.5 Geological applications

On their own, echo sounder and side scan sonar surveys do not provide the basis for any but a very tentative interpretation of seabed geology though they can give a good indication of the variability of seabed materials. When used in conjunction with sampling and drilling, sounder and sonar surveys can be used to produce a map showing distribution of seabed sediment types and the locations of outcropping solid rock wherever such crops have been scanned by the sonar. Echo sounder records and sonographs give qualitative information on both roughness and hardness as well as quantitative data on the shape of bottom features. Hardness variation is indicated by variation in echo strength as printed out on sounder and sonar traces. A very important function of both sounder and sonar surveys is to locate specific targets for subsequent sampling, shallow drilling, or investigation by photography or submersible. For example, they may be required to locate outcrops of solid rocks in surveys aiming to investigate pre-Quaternary geology, see figure 3/5, or under other circumstances to locate areas of possible gravel deposition in surveys aiming to explore for commercially exploitable quantities of this particular bulk mineral.

Where seabed is characterised by extensive areas of rock outcrop, sonographs can be used to determine the strike of rock formations and in some cases pictures are of sufficient quality for faults and folds to be detected and mapped, see figure 3/9. Such dramatic examples apart, it may be said that few topographic features of the seabed have no geological relevance, no matter how insignificant they may at first appear, unless of course they are man-made, such as dredged channels or dumps of spoil. The art of interpreting seabed topography in terms of geological features, like that of interpreting land topography in similar terms, is best acquired with experience.

This is also true of the methods used to classify and interpret the patterns seen on sonographs in terms of seabed surficial sediment distribution. Distinctive patterns are associated with gravel banks and furrows, sand waves, banks, ribbons and ripples and clay bottoms. Featureless zones on a sonograph are generally associated with areas of mud bottom. Some of these patterns are shown in figure 3/10.

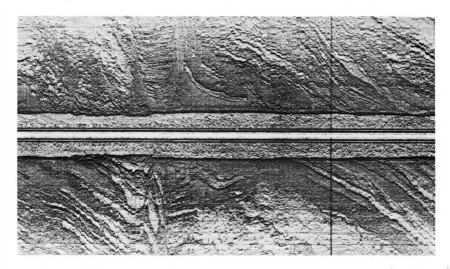

Figure 3/9   Klein Hydroscan record from an area close to the Berwickshire coast, Scotland showing exposed folding and faulting structures. (IGS record.)

Figure 3/10 Klein Hydroscan record showing a variety of patterns associated with different types of seabed sediments. Dark patches are of sand, lighter areas are of silts and muds.
(IGS record.)

A useful interpretation technique is to subdivide the area of investigation into domains of similar morphological type as deduced from sounder and sonar records and to sample the various domains to check identification and the consistency of association between morphology and sediment type. This method usually permits preparation of a fairly reliable map of surficial sediment distribution but care must be taken to allow for the effects of varying sea depths and the presence or absence of currents, as these factors also strongly influence seabed morphology. The better a survey is controlled by carefully collected and analysed samples, the more reliable the final sediment map will be.

3.6 Engineering applications

Engineering applications, particularly of side scan sonar investigations, not directly associated with geological studies are briefly mentioned here, though strictly outside the scope of this book.

It is not known how many wrecks are distributed around the world's continental shelves, but in the North Sea for example about 6000 are known, and many more are expected to exist, but have not been located. The majority do not represent a navigational hazard, or did not in the past, but present day navigational requirements are very different to those of a few decades ago, with tankers drawing up to 20m and oil production platforms requiring towing routes cleared for 80m draughts. The side scan sonar is a very useful instrument for detecting wrecks as well as any other seabed obstruction, see figure 3/11, and has wide application in this field.

Other applications include mapping the courses of uncovered pipelines and cables, see figure 3/12, detection of gas leaks from pipelines, the study of sediment movement and scour around oil rigs and platforms, and the detection and monitoring of dredging activities.

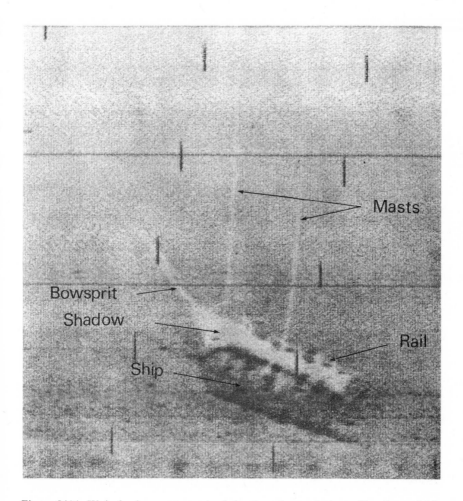

Figure 3/11 Klein hydroscan record of the American schooner *Hamilton* which sank in Lake Ontario during the war of 1812.
(Reprinted with permission of Klein Associates, the Royal Ontario Museum and the Canadian Centre for Inland Waters.)

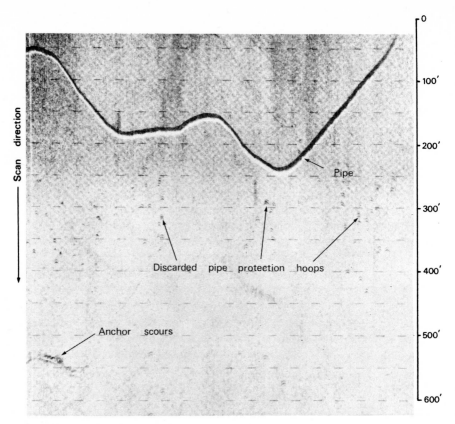

Figure 3/12 UDI AS Series side scan sonar record showing a pipe sitting on the seabed causing a strong acoustic shadow. The small semicircular echoes are from bands used to protect the ends of pipe sections and discarded during pipe welding. Anchor scours from the laybarge anchors are also clearly visible.
(UDI record.)

As the techniques become more refined an even wider range of applications is likely to develop. There have been dramatic improvements in instrumentation over the past ten years and such progress is likely to continue so long as there is commercial pressure for an improved means of viewing the sea floor, its natural condition and the results of man's activity there.

4 Continuous Sub-bottom Seismic Reflection Profiling

4.1 Introduction

Seismic methods are those which depend on the generation and detection of acoustic waves. As so defined, an echo sounder is a simple seismic profiling instrument. Continuous seismic reflection profiling techniques have developed directly as an extension of the echo sounding technique. An echo sounder, particularly if operated at a relatively low frequency in shallow water, will under certain geological conditions record echoes from beneath the seabed; for example, if a hard rock seabed is covered in places by a layer of soft mud then it is quite usual for both the seabed and the underlying boundary between mud and rock to be traceable on the echo sounding record. An example of an echo sounder trace showing penetration to a sub-bottom rock layer is featured in figure 4/1.

The value of the sub-bottom profiling technique has become increasingly apparent to geologists as instruments have been developed capable of recording echoes from layers, not just a few metres beneath the seabed, but from depths of tens, then hundreds and eventually a few thousand metres; increased penetration being achieved by development of high energy acoustic sources and more efficient signal detection and recording systems. Over the past twenty years, many of the great advances which have been made in expanding our knowledge of continental shelf geology have depended more on improvements in the techniques of seismic profiling, than on the development of any other method. More recently the method has become increasingly important in engineering studies and site investigations for such structures as oil production platforms, pipelines, jetties and tunnels. Another important use has been the mapping of structures in areas undergoing development of offshore sub-surface mining. Seismic methods using multi-channel digitally recorded data used in hydrocarbon exploration are discussed in the following chapter. These are closely related to the shallower penetration techniques discussed here, but in hydrocarbon exploration deeper penetration is essential, and this can only be achieved by a marked increase in the complexity of signal processing applied and use of higher energy acoustic sources. There is also a very large cost difference. A good single channel reflection profiling system can be mobilised on ship at a capital cost of a few tens of thousands of pounds. Mobilisation of an equivalent deep exploration system would involve expenditure of a few hundred thousand pounds.

4.2 Basic principles

The basis of all reflection seismic methods is to initiate a pulse of sound at a source point and to determine, at the same or another point close to the first, the time interval between initiation of the pulse and the reception of sound wavelets which have been reflected from discontinuities in the transmitting media; water, sediment and rock. In marine exploration we are concerned primarily with pressure waves

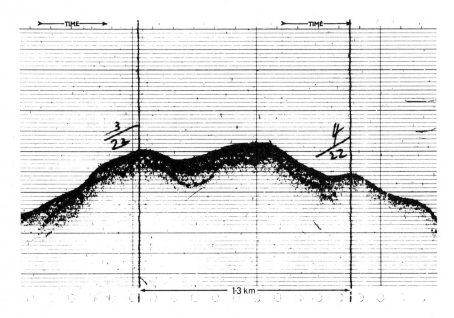

Figure 4/1   Echo sounder record showing penetration to a sub-bottom rock layer.
(IGS record.)

and the velocity, v, of such a wave is given by the equation:

$$v = \sqrt{\frac{(k + \tfrac{4}{3} n)}{\rho}}$$

where k is the bulk modulus,
        n is the shear modulus,
and ρ is the density of the medium.

Hard rigid rock materials have high seismic velocities, soft plastic rock materials have low seismic velocities. Density variations affect seismic velocities less significantly than does rigidity but it is nevertheless a general empirical rule that seismic velocity increases in step with increased density in rocks of similar type. Some examples of seismic velocities are as follows:

Type of material	Velocity
Air	330m/s
Water	1490m/s
Glacial moraine	1600-2700m/s
Limestone	3500-6500m/s
Granite	4600-7000m/s

Reflection of seismic waves takes place at a boundary between layers of contrasting acoustic impedence (seismic velocity X density) the reflection strength depending on the impedence contrast. As can be deduced from the above table of velocities there is a large impedence contrast at the sea's surface, thus all reflection methods must take account of reflection from sea surface as well as those from the sea bed and sub-surface layers. The sea surface is said to have a high reflection coefficient.

An understanding of the definition of reflection coefficient is important. If a sound wave passing through a medium of impedence $\rho_1 v_1$ is incident (at normal incidence) with a medium $\rho_2 v_2$ some energy will be reflected and some transmitted. The reflection co-efficient is given by:

$$C_R = \frac{A_R}{A_I} = \frac{\rho_2 v_2 - \rho_1 v_1}{\rho_2 v_2 + \rho_1 v_1}$$

where A_I is the amplitude of the incident wave
and A_R the amplitude of the reflected wave.

If a wave is reflected from a boundary such that $\rho_1 v_1 > \rho_2 v_2$, for example at sea surface, the coefficient C_R is negative. An important effect of reflection from a boundary having a negative reflection coefficient is that the phase of the reflected wave is inverted. It is not unusual to encounter negative reflection coefficients associated with boundaries in sedimentary sequences though generally impedance increases with depth of burial. Special cases will be discussed below.

4.3 Seismic profiling equipment

The principal components of a seismic profiling system shown in figure 4/2 are almost identical to those of an echo-sounding system as shown in figure 3/2, also, the method of recording data to form a sectional picture of the seabed and underlying rock layers is essentially the same as that described for the echo-sounder, see figure 3/3. A wide range of types of equipment is available and there are many possible combinations of component parts thus allowing great flexibility in adaption of the technique to meet the specific requirements of any particular exploration problem. The user must carefully define the zone of penetration he needs to investigate as well as the required resolution of the system. Sometimes it is important to detect structure in thin layering of less than one metre separation; sometimes it is more important to penetrate a few hundred metres to an important reflector. Seldom is it possible to obtain both these objectives using the same system. To cover the range of possible systems it is convenient to discuss separately the various types available of acoustic source transducer, receiving transducer and recording instrumentation. Deployment of complete systems can then be discussed in terms of gear required for ship installation and towing.

4.3.1 Acoustic sources

Numerous devices have been developed for production of an intense short burst of sound in water at a rapid repetition rate and low running cost. The earliest seismic source for deep exploration work was the high explosive charge, but chemical explosives have never up till now been developed into a suitable system for continuous reflection profiling. Most profiling sources depend on conversion of electrical energy into acoustic energy, though also in common use are pneumatic devices powered by compressed air but triggered electrically. All sources to be discussed here are capable of firing rates of one 'pop' per second or

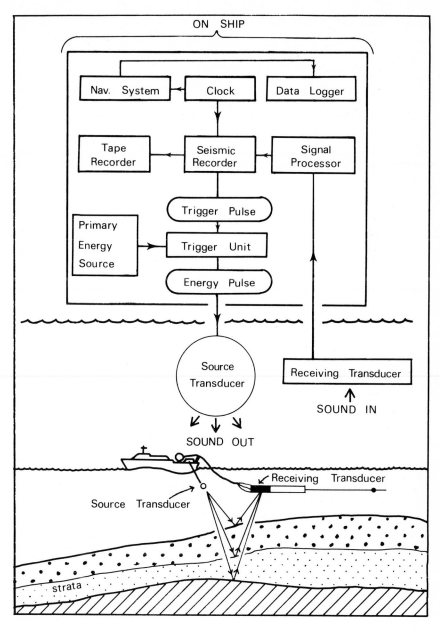

Figure 4/2 Block diagram showing the main components of a continuous sub-bottom seismic profiling system.

faster, though slower firing rates are often used, particularly when deeper penetration is required than one second two-way reflection time (about 1km of sediment penetration). Many devices will operate over a range of energy or power levels and usually only produce maximum power at slow firing rates. No single source is ideal for every possible requirement and all have deficiencies and limitations. The aim is to produce an acoustic pulse, or wavelet of sound, having high energy and short duration. For highest resolution of thin layering, cross-bedding in sediments, detection of single boulders etc, a high frequency pulse must be used. For penetration to depths of a few hundred metres, only sources which provide relatively low frequency energy are likely to be successful. In figure 4/3 the relationship is plotted between resolution and pulse-length, and more approximately, between resolution and source frequency, assuming a near ideal pulse of one cycle length. Thus it can be seen that for very high resolution work a pulse length of less than 1ms is required, and the source must have a dominant frequency of about 1kHz. If increased penetration is required then it is usually necessary to use a lower frequency source, partly because it is possible to manufacture more powerful sources which emit lower frequencies and partly because as sound waves pass through a rock or sedimentary layer then this layer will absorb higher frequency acoustic energy more effectively than lower frequency energy. Attenuation of sound in rocks depends on what is called the quality factor Q, which is the amplitude ratio of energy transmitted per cycle to the energy dissipated. The Q of rock describes its 'quality' as a transmitter of sound and is therefore inversely proportional to the

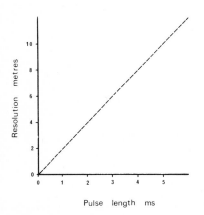
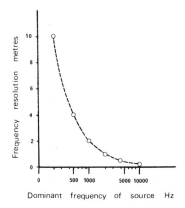

Figure 4/3   Graphs of, on the left, the relationship between seismic resolution and pulse-length and, on the right, the relationship between resolution and source frequency for a 1-cycle pulse.

absorption Z, which is usually quoted in decibels (dB) per wavelength. The relationship between these parameters is given by:

$$Z = \frac{27 \cdot 3}{Q} \ dB$$

If we assume an absorption of 0.5dB per wavelength (Q = 55), a reasonable figure for sediments close to the seabed, then in figure 4/4 we see, for a range of seismic frequencies, the effect of absorption on the amplitude of a sound wave completely reflected at an interface, over a range of depths beneath seabed.

In this diagram we are comparing the amplitudes of a sound wave as it enters the rock layer at seabed with that as it is emitted from the seabed along its reflected path, assuming that attenuation is due to absorption alone. Velocity is assumed to be 2km/s. The decibel scale is a logarithmic scale such that an attenuation of 1dB = $20.\log_{10}$ (attenuated amplitude/original amplitude); thus, an attenuation of 20dB is equivalent to a reduction in amplitude of seismic wave to one tenth its original value; a 40dB attenuation is a reduction to one hundredth; a 60dB attenuation to one thousandth. This example underlines the fact that it is seldom possible to obtain both very high resolution, by using a high frequency source, as well as deep

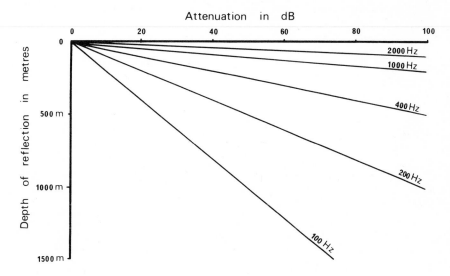

Figure 4/4 Attenuation of seismic waves for an absorption of 0.5dB per wavelength plotted as the effect of absorption on the amplitude of a sound wave reflected at a range of depths 0–1500m.

penetration. Increasing the energy output from a source by a factor of two has little effect as a means of obtaining greater penetration when compared with the effect of lowering the frequency by the same factor.

Power outputs from profiling sources are usually quoted either in terms of joules (energy), or kilowatts (peak power). Low energy devices such as echo sounders and pinger probes are generally quoted in commercial

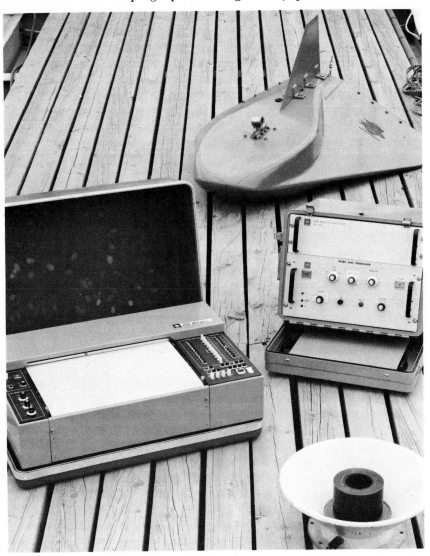

Figure 4/5   An Edo Western 515 pinger system.
(Photo: Polytechnic Engineering Ltd.)

literature in terms of kilowatt power whereas high energy devices such as sparkers and boomers are usually quoted in terms of joules per pulse. The joule being an energy unit and the watt a unit of power it can be difficult to make direct comparison between sources differently defined. However, to illustrate how large these differences can be let us compare a commercially available pinger probe source with a 1000 joule sparker source, remembering that 1 watt = 1 joule per second:

Pinger source: 10,000 watts peak power, 0.5ms electrical pulse length, approximate energy per pulse = 5 joules.

Sparker source: 1000 joules energy per pulse, 1.0ms electrical pulse length, approximate peak power = 1,000,000 watts.

We can now discuss in more detail a number of types of acoustic source currently in use in sub-bottom profiling systems. The transducer most similar to that used in echo sounders is simply a low frequency piezoelectric or magnetostrictive transducer which, to distinguish it from an echo sounder transducer, is often called a pinger probe. The emitted pulse is tuned to a narrow frequency band and is usually quoted as having a discrete frequency in the band 2-8kHz. For example, the Edo Western Model 515 uses a dual frequency transducer which can be operated at either 7.0 or 3.5kHz. Pinger probe sources, like echo sounder transducers, are usually designed to emit a focused beam of sound, with a beam width in the range 30-50° being common. Thus the source must be pointed at the seabed, and the receiver must be close to the transmitter (sometimes the same transducers are used). Furthermore, the towing vehicle must be stable in all sea states in which it is required to obtain good data. A limited beam width along with short pulse length, ensures the required high potential resolving power of such a source. A photograph of a complete pinger system is shown in figure 4/5 and in figure 4/6 an example of recording made off the coast

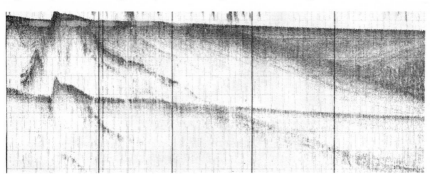

Figure 4/6   An Edo Western 515 pinger record obtained in the Firth of Forth. Horizontal scale lines are equivalent to approximately 12m thickness in sediment.
(IGS record.)

of Scotland illustrates how thin dipping layers of less than one metre separation can easily be resolved. On the same record penetration to identifiable horizons exceeds 50m but it should be noted that the record was obtained in an area where geological conditions are more than usually favourable for obtaining good quality pinger data.

For reasons of cost, weight and other design restrictions, pinger transducers can handle only fairly low energy pulses. A similar type of source, capable of emitting a much higher energy pulse as well as operation across a lower band of frequencies is the device known as a boomer. A wide range of boomers is marketed, designed to operate at energy levels from less than 100 joules up to 5000 joules (per pulse) and within the frequency range 200Hz-15,000Hz. The acoustic output is, however, broad band when compared with that of the pinger thus allowing more flexibility in signal processing together with a combination of good resolution and moderate penetration. It is the source most commonly used in high resolution profiling where the target range is beyond that which can be investigated using a pinger probe. It is most commonly operated in the range 100-500 joules. The design criteria demand a simple lightweight device which will produce a non-ringing half cycle pressure pulse. Operation involves conversion of the energy of an electrical pulse into acoustic energy through a powerful half-cycle deflection of a stiff metal plate. A coil of wire is mounted in a rigid bed of non-conducting material against the face of a flat aluminium plate. A high voltage, short duration, discharge from a bank of capacitors activates the coil, and eddy currents are set up in the metal plate. The magnetic field thus produced in the plate opposes that in the coil and the plate is repelled to produce a sharp powerful deflection. The pressure pulse produced by a Huntec ED10 boomer is shown in figure 4/7. It can be seen that this pulse is approximately 0.1ms duration, and frequency analysis shows a broad energy spectrum from below 2kHz up to 15kHz.

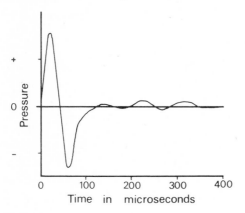

Figure 4/7
Pressure pulse
produced by
Huntec ED10 boomer.

Figure 4/8 A Huntec ED10 boomer transducer mounted in a catamaran. Above: underwater photograph. Below: preparation for launch. (Photos: Huntec.)

In figure 4/8 a boomer is seen mounted on a catamaran for shallow water survey work at slow speeds of up to 4 knots. Boomers can be mounted in towing fish for work in deep water and for towing at speeds of up to 8-10 knots. The ED10 boomer transducer, alone, weighs 80kg. Figure 4/9 shows a record obtained using an ED10 in a North American lake illustrating the combination of good resolution with high penetration which can be achieved through a variety of sediment types and into bedrock. The boomer source has other features which make it comparable with a pinger probe; these include its directionality, the energy being concentrated into a beam of sound, and the fact that it can be designed so that upward radiation is reduced to a very low level, thus eliminating 'ghost' reflections from sea surface. Although large boomers capable of handling electrical pulses of up to 5000 joules are available, these are very heavy and not easily handled at sea. Furthermore, the boomer is a relatively inefficient device in converting electrical energy into acoustic energy. When penetration beyond that obtainable using a high resolution boomer is required, it is common practice to employ an omnidirectional (as opposed to focused) source such as a sparker or an air-gun.

Sparker systems are available which operate in the 100-200 kilojoule range, but these are used as sources with multi-channel seismic systems, not in continuous sub-bottom profiling. Sparker sources used in profiling are usually limited to the range 200-10,000 joules and generally there is little advantage to be gained in a further increase in energy beyond 3000 joules. Investigations using sparkers as sources for single channel profiling systems are usually confined to the penetration range 0-1.0 second two-way reflection time below seabed. The principle is very simple; during the period between pulses a bank of condensers, similar to that used to drive a boomer transducer, is charged to a few thousand volts, usually in the range 4-10kv, then a trigger circuit is operated which discharges this stored electrical energy directly into the sea through a number of metal electrodes. For safety, it is normal practice to tow an earth plate close to the array of sparker electrodes. At the tip of each electrode a small explosion occurs with the production of a bubble which oscillates then collapses after a few milliseconds, the period and frequency of oscillation depending on the size of the discharge. Early systems used a single sparker tip, or in some cases three or nine tip arrays. Modern systems are more usually equipped with multi-electrode arrays which discharge through between 100-1000 tips. The advantages of the multi-tip assembly include shorter pulse lengths for an equivalent discharge energy as well as an increase in peak pressure, i.e. the amplitude of outgoing acoustic wave. Sparkers are capable of giving an excellent pulse for work in the

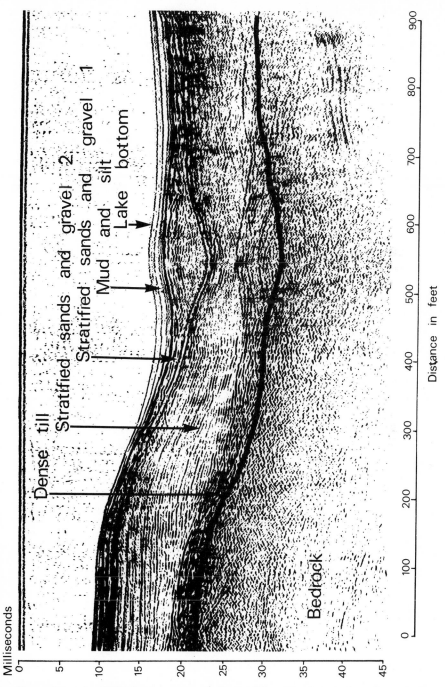

Figure 4/9 An ED10 boomer record obtained in a North American Great Lake. (Huntec record.)

200-1000Hz frequency range and can provide a most versatile source for all types of reconnaissance investigation. One considerable advantage is that the pulse can be easily altered without any break in survey by simple adjustment of energy output per discharge and by changing the towing depth of the source array. Thus on-line adjustments can be made to suit varying water depths, geological conditions, or varying depth of target horizons. In figure 4/10 an EGG 3-tip spark array is shown alongside a simple 200-tip disposable array and in figure 4/11 is shown a complete sparker profiling system, capable of delivering 1000 joule pulses every second. A sparker record obtained concurrently with the pinger record (see figure 4/6) is shown in figure 4/12, and the increase in penetration at the expense of resolution is easily seen by comparison of these records.

One other seismic source in use with profiling systems, though it is generally more suitable for deep water profiling than for work in continental shelf depths, is the air-gun. The main disadvantage of this source as a profiler source for shelf exploration is that, for most purposes, it produces an excessively long pulse though in recent years modifications to early designs have significantly improved this aspect of performance. Air-guns are discussed in more detail in the following chapter.

4.3.2 Acoustic receivers

When a pinger probe is used as an acoustic source, the same transducer may be used both as a transmitting and receiving device, in much the same way as is commonly the case with echo sounders. With all other sources, a separate receiving transducer or array of transducers must be towed in the water. The design of this acoustic receiver is determined largely by the principal characteristics of the source, the noise characteristics of the sea environment, and the particular exploration objective. To minimise the generation of noise as the receiver is towed through water, sensing elements are usually mounted inside a flexible tube, the resulting receiver assembly being called a seismic streamer or eel.

Individual sensing units are called hydrophone elements. A hydrophone element detects pressure variation in the sea water through which it is towed. Magnetostrictive phones can be used, but piezoelectric elements are much more common. A primary aim is to obtain high sensitivity to pressure variation, with minimum sensitivity to local accelerations. This contrasts with geophones, land seismic detectors, which are designed as acceleration detecting devices. A

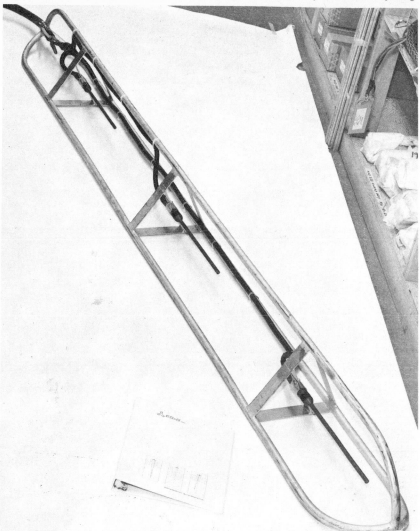

Figure 4/10 Sparker acoustic sources; EGG 3-tip array and an IGS 200-tip array. (Photos: FEP and IGS.)

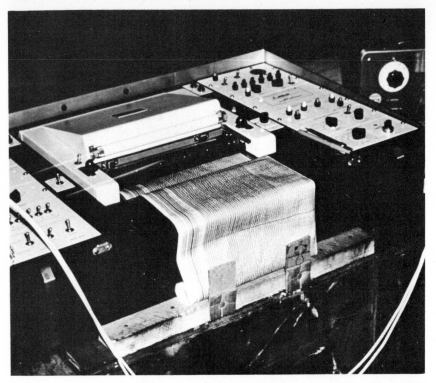

Figure 4/11 A sparker profiling system manufactured by EGG. Above: the recorder. Below: power supplies and capacitor banks. (Photos: IGS.)

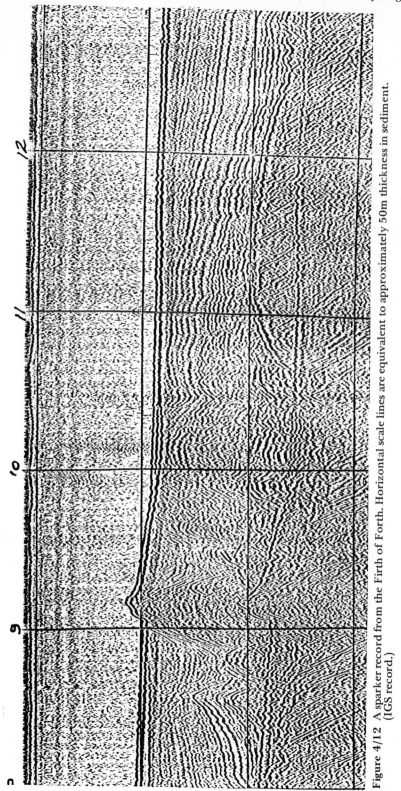

Figure 4/12 A sparker record from the Firth of Forth. Horizontal scale lines are equivalent to approximately 50m thickness in sediment. (IGS record.)

typical hydrophone element is shown in figure 4/13, and is seen to be about 2.7cm diameter. For high resolution surveys, it is necessary for the length of the active section of the streamer, that is, the part of the streamer containing hydrophone elements, to be only a few metres long. For some work a single element detector is used mounted in a short eel. For deep penetration and good response to relatively low frequencies, down to 100Hz, it is desirable to have an active section of ten metres or more in length, which may contain as many as fifty hydrophone elements.

Such an assembly is sometimes called an array hydrophone streamer. Streamers are carefully balanced to be neutrally buoyant so that noise generated by flow of water over the surface of the eel during towing is minimised. To achieve this, the tube is filled with oil and small adjustments of buoyancy are made by altering the oil content. Array streamers are usually fitted with pre-amplifiers which sum the signals output from all the elements and transmit the summed amplified signals to the shipboard recording system. The main advantage of an array detector over a single element detector is that noise generated by the ship is of fairly low frequency and, being transmitted along the length of the streamer, can be to a large extent cancelled by summation of the group of out-of-phase signals. By contrast, sound reflected from the seabed and deeper levels arrives at the streamer along ray paths approximately normal to the streamer axis; thus these signals are in phase and on summation add. Another potential source of noise is that caused by transmission and coupling of the ship's motion and vibration to the streamer elements through the tow cable. It is because horizontal accelerations are transmitted to the cable as the ship heaves and pitches that a low sensitivity to horizontal acceleration is an important feature of the design of a hydrophone element; particularly accelerations axial to the streamer.

One method whereby acceleration cancellation is achieved is by mounting dual crystals in each hydrophone element such that axial acceleration generates equal but opposite phase signals which when summed effectively cancel each other out. Advances in the design of hydrophone elements and streamers made over the past few years have considerably increased the efficiency and helped to reduce the cost of seismic profiling operations. It is now possible to acquire good quality data at higher survey speeds of up to 10 knots than previously possible and under worse weather conditions.

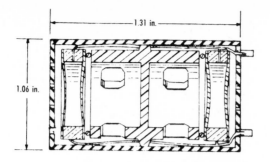

1A.   Cross section of complete hydrophone assembly

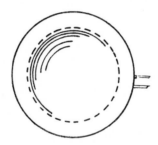

1B.   Cross section of hydrophone element

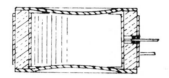

1C.   Top view of hydrophone element

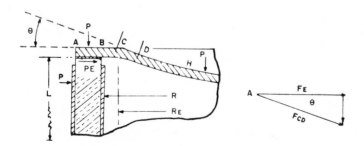

1D.   Enlarged view of diaphragm-crystal assembly

Figure 4/13 A Multidyne hydrophone.
(Published with permission of Seismic Engineering Company.)

4.3.3 Recording systems

On ships, seismic records are displayed on a facsimile recorder using a format similar to that of the echo sounder, and the description of the echo sounder recording system in chapter 3 is generally applicable to the seismic recorder. Indeed, on many ships identical oceanographic recorders are used for both echo sounding and seismic profiling. Referring again to figure 3/2, for sub-bottom profiling work the receiving amplifier and signal processor would need to be of special design to handle the wide band of frequencies, and the display unit and timing mechanism would need to be operable over a wide range of sweep speeds and trigger pulse intervals. A most desirable feature of some modern recorders is that analogue magnetic tape recordings can be made in parallel with presentation of the visual display, thus allowing replay of seismic sections using a variety of signal processing techniques. A brief description of a typical graphic recorder is given below, see figure 4/14.

This particular recorder uses a digital system of direct stylus drive; many other recorders use a synchronous motor with phase-locked loop speed control. The digital drive allows reproduction of tape recordings at faster speeds than set during original recording, an important factor if much replay work is envisaged. Also, the stylus sweep can be

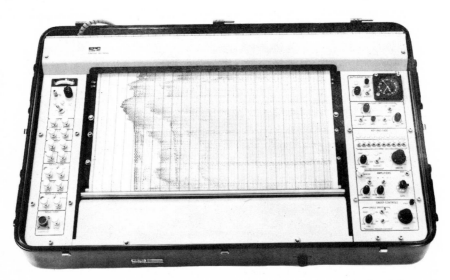

Figure 4/14 An EPC Model 4100 graphic recorder with seabed sub-bottom seismic recording.
(Photo: Euro Electronic Instruments Ltd.)

programmed to be driven externally by non-linear clock frequencies thus allowing, for example, slant range correction during display of side-scan sonar records. A time base is provided from an internal crystal oscillator. In common with most modern recorders, the sytem does not include an internal filter, but an interface connector is provided on the front panel into which an external filter can be plugged. This is inserted after the first stage of amplification of the input signal. An active band-pass filter unit is used when the instrument is being operated as a seismic recorder, the pass band being selected to suit the type of acoustic source in use and the resolution/penetration requirement of the survey, as well as to minimise the effect of any major source of noise. In some seismic profiling work, band-pass filtering is the only form of signal processing applied, although recording systems often provide for both time varying gain (TVG) and/or automatic gain control (AGC). A TVG facility is important, for example, when pinger probes are used, and this is often constructed such that the TVG can be triggered by the first reflection from seabed thus allowing the operator to apply TVG across the time interval of main interest, that between seabed and the deepest detectable reflector.

4.3.4 Towing gear

For high resolution work, seismic source and receiver are usually towed close to the sea surface, it being desirable not to have sea-surface reflections interfering with actual reflections from beneath the seabed. With pinger probes this is not an appreciable problem because of the directional selectivity of both source and receiver, but with more powerful, omnidirectional sources such as sparkers operated in conjunction with array receivers, towing depth should be carefully selected and controlled for best results. As a general guide, it is best to tow both transmitter and receiver within one-half wavelength below sea surface of the dominant frequency being recorded. Thus if a system is operated with filters set for the band 300-800Hz with a dominant frequency of approximately 500Hz, then a tow depth of 1.5 metres or less would be recommended. However, towing this close to the sea surface can, in choppy seas, lead to the generation of noise along the hydrophone array, and a deeper tow may be necessary even though this lengthens the composite direct and reflected pulse length. In predicting the effects of differing depths of tow on the shape of a received signal, it should be remembered that sound waves undergo a half-wavelength phase shift on reflection from the sea surface.

For some applications, particularly where it is necessary to elucidate details of complex geological structure close to the seabed (say in the top 20 metres of rocks and sediments) it can be desirable to operate a source and receiver system fitted into a deep-tow body. Advantages include: closer proximity to the seabed, giving increased resolution and in some cases greater penetration; the relative acoustic quietness at depth and therefore better signal to noise ratio; and decoupling of the towed body from ship and wave motion which gives a high resolution display undistorted by superimposition of the effects of vertical motion in the transmitter and receiver. Additionally, in the case of a deep-towed sparker, the effect of high hydrostatic pressure is to shorten the sparker pulse and increase the frequency of the sound wavelet to give for most purposes, a better acoustic pulse. At the time of writing, two systems are known to the authors to be in commercial use; a deep tow boomer system developed by Huntec of Canada, and a deep tow sparker system developed by the Nova Scotia Research Foundation, again of Canada. Other organisations are known to be developing tools of similar type. One of the main difficulties is that of obtaining an efficient and easily-operated handling system and another is that of providing automatic correction for change of depth of the transducer fish as results are displayed at the final recording stage. An example record shown in figure 4/15 was obtained using a deep tow sparker in an area of the North Sea.

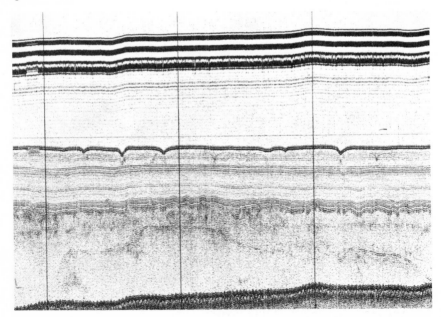

Figure 4/15 A deep tow sparker record from the North Sea using Nova Scotia Research Foundation (NSRF) equipment.
(IGS record.)

4.4 Analysis of profile data

A seismic record, as displayed by a graphic recorder, presents a picture of geological structure which is spatially distorted. This distortion arises because the recording system displays time-related not space-related measurements: along one axis ship travel time, along the other two-way travel time of sound which has passed through media of varying seismic velocity. To illustrate the methods of analysing a profile section the records shown in figure 4/16 will be used; these are short lengths of two intersecting profiles, the relative location of which is shown in the inset plan. The first job is that of identification of important horizons for measurement and structural interpretation; at the same time multiple reflections, side-reflections, single-point reflections and diffraction patterns must be identified, so that these are not misinterpreted as real reflections. Direct reflections from geological structure below the shot-point are termed primary 'real' events, whereas events seen on the seismic section which are caused by either multiple reflections or non-vertical reflections are usually termed secondary, 'spurious' events. Figure 4/17 shows how multiple reflections can lead to the development on sections of a wide variety of secondary events. In figure 4/16, multiple events are marked; type 1 is the first seabed multiple, types 2 and 3 are events which derive from the combination of a primary reflection from a rock layer (top of till) followed by a seabed multiple along with a seabed multiple reflection followed by a primary reflection from the same rock layer, and also type 6, single-point reflection events caused by echoing from places where hard rock layers crop out at rockhead beneath superficial sediments.

In figure 4/17 it should be noted that a multiple reflection from a sloping horizon, type 5, leads to the development of a secondary event which has a slope which is twice the slope of the primary event. This effect can be used to distinguish between primary and secondary events. Also, single-point reflection events, when corrected for scale distortion, form hyperbolae with corrected dips approaching 45°. Many more types of secondary and spurious events exist than shown in figure 4/17, where only the simplest and most common are illustrated. The ability to recognise such events is to a large extent acquired through practical experience of interpretation.

Recognition of secondary events allows the selection, or 'picking' of the main primary events on the section. Some such events are marked in figure 4/16. Although it may not be necessary to replot a complete survey in a scale-corrected form, it is useful to produce a number of

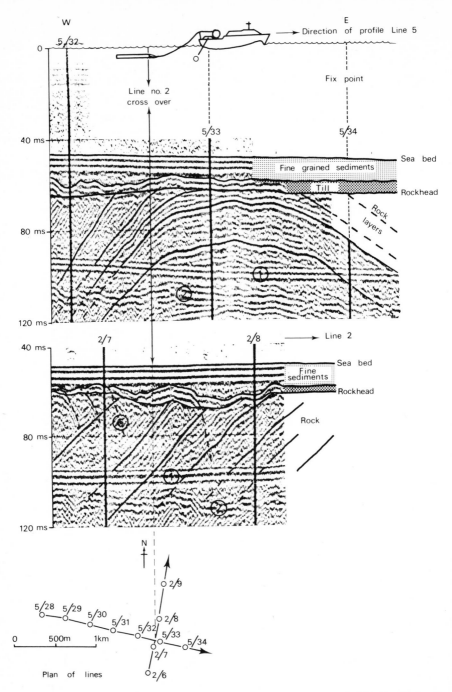

Figure 4/16 Two intersecting sparker profiles the positional relationships of which are shown in the plan inset.

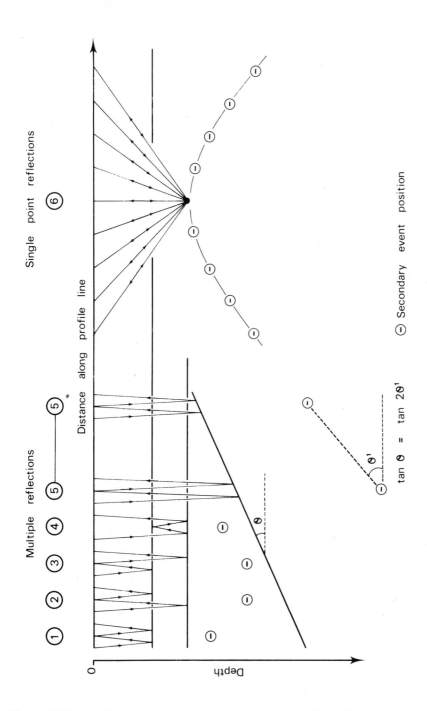

Figure 4/17 Secondary seismic events due to multiple and single point reflections.

corrected sections to illustrate the effect of scale distortion. This is often accomplished using computer systems: a picked section is digitised on a digitising table, the data then being processed by computer and the results plotted on a drum plotter. To convert the time section into an accurate depth section precise values of seismic velocity would be required for application throughout the section. These values are seldom available in detail but an approximate conversion is still very useful. Using the upper section in figure 4/16 and assuming the following velocities; water 1490m/s, superficial deposits 1800m/s, rock layers 4000m/s, a plot has been prepared in figure 4/18 using equal horizontal and vertical scales. If the detail of a complete graphic section is not required a map can be plotted showing values of depth to important horizons, so that these may be contoured to give, for example, a map of thickness of superficial sediments over rockhead. Dip and strike values can be computed at selected points for inclusion on a structural map.

Calculation of dip in dipping formations again depends on an assumed value for seismic velocity; calculation of strike is independent of such an assumption. In figure 4/16, in the upper profile section between fix 5/32 and the intercept of line 2 with line 5, beds dip to the west at an angle on the record of approximately 50°. This corresponds to an angle of between 25° and 30° on the scale corrected section in figure 4/18. To make a direct calculation of scale corrected dip along profile the following measurements must be made:

D = surveyed distance between fixes from plotting sheet (in metres),

d = distance between fixes as measured on record (cm),

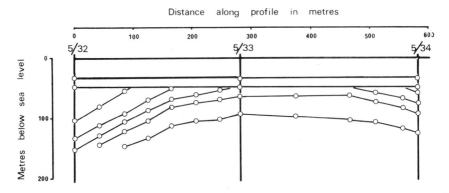

Figure 4/18 Plot of main seismic reflectors in line 5 (figure 4/16) using equal horizontal and vertical scales and correcting for velocity variation with depth.

t = time interval between time lines (seconds),
h = distance between time lines measured on record (cm),
v = seismic velocity (m/s), derived from another source or assumed,
A = dip angle measured on record (degrees).

B, the corrected dip angle in degrees is then given by:

$$\tan B = \tan A \ \frac{vtd}{2Dh}$$

Using measurements made on original records, the same example as above; $D = 280m$, $d = 6.8cm$, $t = 0.04sec$, $h = 4.3cm$, $v = 4000m/s$ and $A = 50°$. The corrected value B is thus calculated to be $28°$. Similarly, the corrected dip on line 2 where it intersects with line 5 is calculated to be $24°$. From these two values of corrected dip and data on headings of the two lines it is possible to calculate true dip and strike.

The corrected dip, B_1, along a line having a bearing C_1 relative to north $(0 \leqslant C_1 \leqslant 360)$ is given by:

$$\tan B_1 = \tan B_t . \cos (C - C_1)$$

where B_t is the true dip and C the bearing of the direction of the true dip, ie at $90°$ to the strike. Thus at an intersection of any two lines:

$$\frac{\tan B_1}{\cos (C - C_1)} = \frac{\tan B_2}{\cos(C - C_2)}$$

which by rearrangement gives:

$$\tan C = \frac{\tan B_2 . \cos C_1 - \tan B_1 . \cos C_2}{\tan B_1 . \sin C_2 - \tan B_2 . \sin C_1} ,$$

In our example, the line headings are line 2; $12°$ and line 5; $103°$, thereby giving, at the line cross-over, a value for C of $63°$. By substitution in the equations:

$$\tan B_t = \frac{\tan B_1}{\cos(C - C_1)} = \frac{\tan B_2}{\cos(C - C_2)} ,$$

we obtain a value for B_t, the true dip, of $35°$.

Analysis of profile records, as well as providing the numerical data described above, includes also the indentification of faults, synclinal and anticlinal axes (such as the anticlinal axis seen in figure 4/16). All the structural observations are compiled onto maps which form a structural basis into which geological sampling data is fed to give finally the cartographic synthesis—a geological map. At this stage a more subjective analysis is often important: this involves the recognition of a quality of the profile records usually termed the seismic character. Seismic character is closely related in many cases to lithology and variation in seismic character allows the interpreter to map rocks of similar character and mark boundaries between areas of differing character. The geological validity of mapping an area into domains of similar character can be tested by sampling or drilling. Much reconnaissance mapping in continental shelf areas depends on this technique. In figure 4/16 the seismic character is typical of an Upper Palaeozoic sequence of rocks with good lithological contrasts in the succession. The rocks of this area are of Westphalian or Namurian age in the Carboniferous.

4.5 Geological applications

Seismic profiling records provide the geologist with a detailed picture of geological structure beneath the seabed. In the previous sections it has been shown that by varying the type of source, receiving and recording systems, options may be selected to increase either resolution or penetration. Often, two systems are operated simultaneously without significant interference; for example pinger probe and sparker records may be obtained concurrently along a profile. Thereby a section may be constructed which displays the main features of the geology of both superficial sediments and the uppermost layers beneath rockhead. The method is so versatile that it is generally accepted as an essential part of all geological investigations on the continental shelf where knowledge is required in the vicinity of the seabed. Even in hydrocarbon exploration, where targets are beyond the normal depth of penetration of continuous profiling, profile data is often invaluable as a means of defining the near surface expression of deeper structure, especially where shallow drilling or seabed sampling can be used to obtain rock specimens for stratigraphical and lithological identification. A number of specific uses of profile data are briefly described below commencing at the seabed and working downwards.

In most parts of any continental shelf some superficial deposits occur. Around Britain, the continental shelf is, except in the extreme south, an area which was glaciated in the Quaternary period and where there now exists a widely varying thickness of unconsolidated glacial and recent sediments. These have been deposited on a rockhead ranging in age from pre-Cambrian to Tertiary. Profile records provide the best means of identifying sites, or areas, where such sediments are locally absent and consolidated pre-Quaternary rocks crop out at the seabed. In geological studies, an outcrop map is usually prepared as a guide to areas where seabed sampling techniques may be applied to the study of pre-Quaternary geology. In figure 4/19 a seismic profile indicates a possible sample site where, in this case Palaeozoic, rocks either crop out or have a very thin cover of sediments.

Knowledge of thickness variations in the uppermost layer of superficial deposits is basic to an understanding of the geology of Quaternary and Recent deposits. Such variations can be mapped using seismic profile data, as indicated in the previous section, though an assumption must be made of seismic velocity in the sediments. If large variations do not occur within distances less than line separation of the survey grid it is possible to produce a series of maps variously termed as follows:

    (i)   Bathymetry, depth to seabed or seabed isobath.
    (ii) Depth to rockhead or rockhead isobath.
    (iii) Superficial sediment thickness or sediment isopach.

Having defined the form of the superficial sediment layer, the seismic profiles can now be used to identify structure and lithological variation within it. Muds, clays, sands, gravels, glacial till, etc, all produce

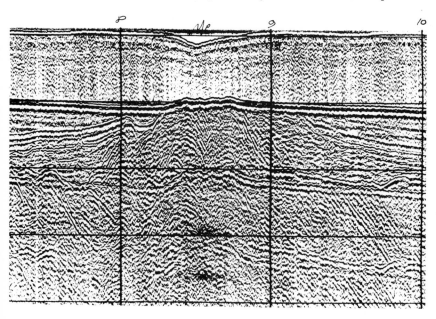

Figure 4/19 Seismic profile showing site where bedrock possibly crops out at seabed. Horizontal scale lines are equivalent to approximately 50m thickness in sediment.
(IGS record.)

characteristic profile records and with adequate control from drilling
or seabed sampling a three dimensional picture can be developed of the
geology of this uppermost layer. Development of this picture, along
with a study of sonar data, also provides the best means of mapping
variation in the type of sediments exposed at the sea floor, assuming
this interpretation is controlled by data from carefully obtained
sediment samples as described in chapter 8. Thus profiles can be used to
elucidate sediment variation on the sea floor, lithology and structure of
the superficial layers, and the physiography of rockhead.

The next main application of the method is to elucidate structure and
lithological variation in layers beneath rockhead using the analytic
methods described in the previous section. True-scale plots of rock
layering can be produced; dip, strike and thickness values calculated,
fault positions located and subcrop positions of marker horizons
mapped, along with the trends of fold axes. The results of all such
measurements are incorporated as structural data on a geological map.

4.6 Engineering applications

Engineering applications of the seismic profiling method are as varied
as the range of engineering endeavours carried out offshore whereby
man-made structures are placed on the seabed, or into layers of rock
and sediment beneath the seabed. A seismic survey is generally
accepted as an essential constituent of any rig, platform or pipeline site
survey. Also, the method is used to explore for, study and monitor a
range of geological hazards which can endanger or add to the costs of
offshore engineering projects. High resolution profiles are used to
explore for man-made objects wherever the requirement is to locate or
re-locate such artifacts, such as buried wrecks, archaeological sites,
buried cables or pipelines. The method is used to study geological
structure in association with offshore underground mining and
tunnelling: examples around Britain include studies made along the
route of the projected Channel Tunnel under the Dover Straits and
studies in areas of projected undersea extensions of coal-mining
operations.

Engineering site surveys have two functions. Firstly they allow the
engineer to avoid, if possible, geological conditions which will make the
engineering project more difficult or more costly. Secondly, they
provide basic information on geological conditions which is necessary
for planning future drilling, piling, dredging, excavation, tunnelling, or
whatever other operations the project demands. In figure 4/20 profiles
with a 1000 joule sparker are shown across an area of varying seabed
conditions near the Orkney Islands, close to where North Sea oil

Shallow seismic profiling.

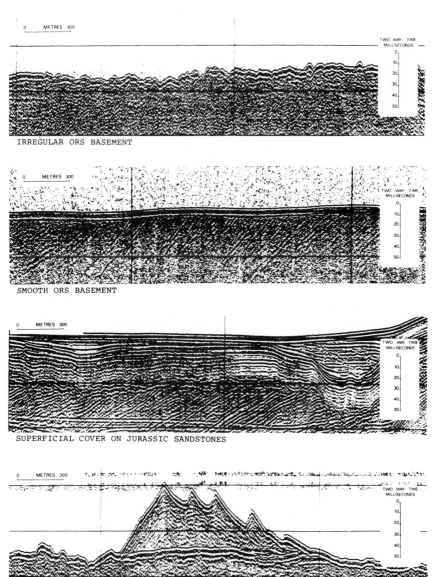

IRREGULAR ORS BASEMENT

SMOOTH ORS BASEMENT

SUPERFICIAL COVER ON JURASSIC SANDSTONES

MOBILE SUPERFICIAL COVER OVER ORS BASEMENT

Figure 4/20 Seismic profiles across a range of seabed conditions in the vicinity of the Orkney Islands.
(IGS record.)

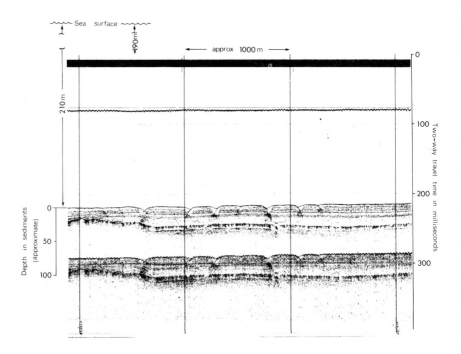

Figure 4/21 Huntec deep tow boomer profile from offshore eastern Canada. (Huntec record.)

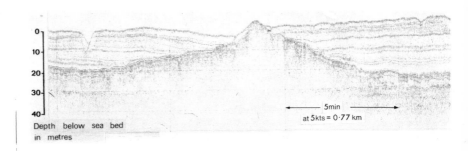

Figure 4/22 NSRF deep tow sparker profile from offshore eastern Canada. (NSRF record.)

pipelines pass. These sections show the type of problem which occurs when a pipeline route needs to be planned. It can be seen that the seabed conditions vary from, on the one hand, thick sediments into which it is a relatively simple task to dredge a channel for pipelaying, to, on the other, rough rocky topography through which it might be necessary to blast a safe channel; a very costly operation. The very large sand waves (underwater sand dunes) shown on this figure present a different problem. These are mobile bodies of sediment and if a pipeline is laid across them large stresses can develop, or in some circumstances spans of the pipe may become unsupported thus leading to a high risk of pipe fracture. In the southern North Sea it has been found necessary to sweep a route through such dune formations prior to the main trenching operation. Obviously a seismic survey is invaluable as a preliminary to any such operation as it gives an indication of the form and base level of the sand bodies. Similarly, it can be seen that the engineering problems of siting a large oil production platform on the seabed are mainly controlled by geological structure in the uppermost 100 metres of sediment and rock layers beneath the platform site. Seismic surveys are used, usually along with test drilling, to establish the requirements for foundation engineering work.

The occurrence of shallow gas in areas being drilled in the Gulf of Mexico has led to a number of major blow-outs, with consequent catastrophic results. Such events have received considerable publicity but indicate the nature of only one of a range of geological hazards which, if not detected and avoided or compensated for, can lead to loss of life, damage to equipment, or at best the loss of valuable time. Modern high resolution seismic profiling techniques go a long way towards providing information about most forms of potential hazards. In figure 4/21 and 4/22, two seismic profiles, both from off eastern Canada, show the presence of pockmarks, seabed depressions thought to be associated with gas release from underlying sediment layers, indicating that these features, well known in the North Sea, represent a geological hazard which may have world-wide occurrence.

5 Deep Reflection and Refraction Seismic Methods

5.1 Introduction

The seismic reflection technique is the basic tool in offshore exploration for hydrocarbons. In the search for offshore oil and gas, other geophysical methods, along with geological sampling and shallow drilling, are of subsidiary importance. It is the analysis of seismic reflection profiles which provides a means of identifying and mapping prospective hydrocarbon traps, and on the basis of these identifications sites are selected for exploration drilling. Furthermore, once oil has been discovered in a reservoir, seismic data are essential to any evaluation of its size and shape and consequent estimate of recoverable reserves. The amount of seismic reflection line shot in the North Sea and adjacent areas since exploration commenced in the early 1960s is in the region of 1,000,000 kilometres. This represents an exploration investment of around £100 million, and this is only one of a large number of actively explored hydrocarbon provinces throughout the world. With such large exploration investment involved, there has been intense commercial pressure and competition to provide ever-improving techniques, better and more cost effective equipment. Considerable funds have been available for research and development with the outcome that the seismic exploration industry is not only a major user of the most advanced electronic and computing techniques but, furthermore, major technological advancements have developed in the industry to find wide application elsewhere.

The seismic refraction method is relatively unimportant in either exploration work or geological mapping; a discussion of the method and some of its applications is nevertheless included as it is still often used in reconnaissance structural studies.

5.2 Basic principles of multi-channel reflection seismic surveying

The basic principles of the seismic reflection method are discussed in section 4.3 of the previous chapter. However, chapter 4 was concerned only with the single channel seismic technique; the seismic method as an extension of echo sounding. Here we are concerned with multi-channel surveying the application of which has made it possible to survey offshore geological structure with sufficient detail, clarity and lack of ambiguity to meet the main requirements of exploration for hydrocarbons. A survey ship tows a hydrophone streamer made up of a number of sections, each of which acts as an independent receiver. Modern streamers are fitted with 24, 48 or 96 such sections and the seismic signal received at each section is separately recorded aboard ship in a digital recording system. In figure 5/1 the reflections from a single horizon are schematically portrayed as received in the first six sections of a 48-section streamer. Let us assume the distance between sections is 50m and that the ship is travelling at 8km/h (approximately 4kts); then if the first shot is at time t_1, a reflection from depth point no. 1 is received in section one of the streamer thence recorded on ship in channel no. 1 of its recording system. The next shot S_2 is timed so that the ship has progressed to a position such that the reflection received in section two of the streamer is from the identical depth point as is received in section one from shot no. S_1. It can be seen that this position is half the interval between sections along the ship's course, that is 25m. The interval between shots $(t_1 - t_2)$ should be set therefore at 11.25sec. Shots S_3 to S_{48} follow at the same interval and successive records are obtained on ship from the common depth point number one (CDP 1) until finally the record from shot no. S_{48} is recorded in channel no. 48 of the recorder. It should be noted that as the ship progresses along course the seismic signal reflected from CDP 1 will have travelled ever increasing distance between shot point and receiving streamer section. However, if this change of geometry is corrected for, it is possible to add together (stack) all 48 seismic records pertaining to CDP 1. This stacking procedure is accomplished in a computer; channel 1, shot point 1 is added to channel 2, shot point 2 etc. It is this process, CDP stacking coupled to other digital computing operations which allows differentiation between primary reflections from geological structure and multiple reflections in sea and rock layers leading to an eventual printout of a seismic section which under most circumstances gives an accurate, detailed and reliable portrayal of rock structure beneath the line of the ship's course.

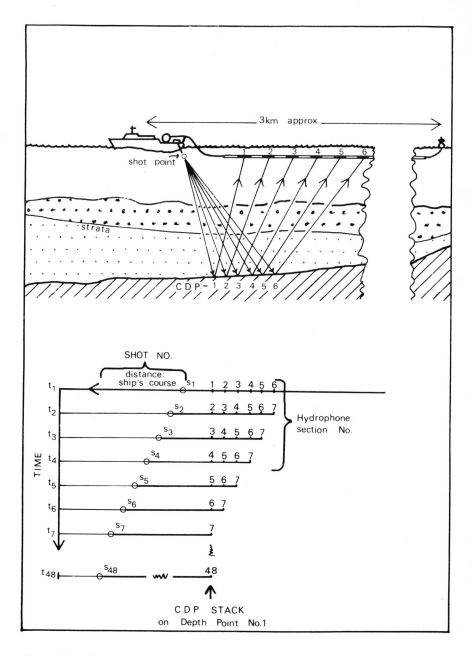

Figure 5/1   Schematic diagram showing the use of multi-channel hydrophone streamers to acquire data which can be common depth point (CDP) stacked.

5.3 Reflection seismic equipment

5.3.1 Marine seismic sources

Acoustic sources for continuous seismic profiling (chapter 4) have been designed to produce intense short duration bursts of sound. Sources for use in multi-channel reflection surveys at sea are designed to have a much higher total acoustic energy output and to include a large part of this energy within the low frequency band between 20 and 100Hz. A major difference between the two types of source is that with the larger sources it is usually a feature of the energy release mechanism that a bubble is formed as a consequence of the initial impulse and that this bubble oscillates to produce a number of secondary seismic impulses.

Early seismic surveys at sea were conducted using conventional chemical explosives as a seismic source. To avoid the generation of bubble pulses, fairly large explosive charges (50-100lb per shot) were used and detonated sufficiently close to the sea surface for the gas bubble to break surface venting into the atmosphere and producing a large water plume. This method is no longer used for a host of reasons: e.g., danger, expense, logistical restriction on operation of ships carrying large quantities of explosives, difficulty of maintaining rapid firing rates. Chemical explosives are still widely used, but as small charges deployed in devices which control or minimise the bubble pulse effect.

There is now available a very wide variety of marine seismic sources, all capable of producing a pulse of seismic energy equivalent to the 50lb TNT charge previously used. A number of the more popular devices will be described briefly, then the nature of the seismic pulses produced by such devices will be compared and discussed.

A widely used non-explosive source is the air gun which functions by discharging a bubble of highly-compressed air into the water. The advantages of using such a device are that compressed air can be safely and continuously supplied by standard compressor equipment on board ship and triggering can be precisely timed by operation of an electrical solenoid within each gun used. Thus it is possible to fire a group of guns simultaneously. The principle of operation is shown in figure 5/2 which is a schematic diagram of the PAR air gun manufactured by Bolt Associates, Inc. High pressure air (say 2000psi) is supplied by a high pressure hose line. During the charging cycle the

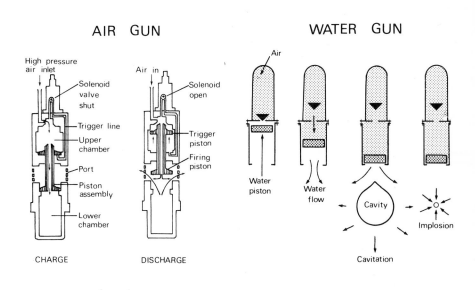

Figure 5/2   Schematic diagram showing the principle of operation of four different non-explosive seismic sources. Top left: PAR Air Gun (registered trade mark of Bolt Associates. The air gun is patented under British Patent Specification Nos. 1,090,363; 1,175,853; and 1,322,927). Top right: MICA Water Gun, courtesy Société pour le Developpement de la Recherche Appliquée (SODERA). Bottom left: VAPORCHOC seismic source (courtesy Compagnie Générale de Géophysique (CGG). Bottom right: FLEXICHOC seismic source (courtesy Géomécanique).

pressure in both upper and lower chambers builds up to line pressure, air from the upper chamber feeding through the centre of the piston assembly to the lower chamber. The piston assembly is held in the downward position because the area of the trigger piston is larger than that of the firing piston. To fire the gun, an electrical pulse opens the solenoid valve allowing a slug of compressed air to travel down the trigger line to the underside of the trigger piston; this lifts the piston assembly, equalising pressure above and below the trigger piston so that the compressed air in the lower chamber drives the piston assembly forcibly upwards releasing the air in the lower chamber into the water. It is this explosion of compressed air which provides the primary seismic pulse. On release of pressure in the lower chamber the piston is rapidly driven back to its sealed position by compressed air from the inlet and during this second part of the cycle the solenoid valve is again closed. The whole piston cycle lasts only about 10ms. The dominant frequency of a pulse generated by an air gun is controlled by the air pressure, the size of the lower chamber, and the depth of operation. For most applications, an array of guns of varying sizes is towed at about 10m below sea surface; these are fired simultaneously to give a pulse which has a broad frequency spectrum and has the main energy concentrated in the initial pressure pulse. If it is impracticable to employ an array of guns, a modified single air gun can be used. One such modification is the Bolt Associates wave-shape kit which bleeds air into the bubble to dampen the effect of bubble collapse and oscillation. A more radical solution is offered by the water gun system designed and marketed by Sodera (Société pour le Developpement de la Recherche Appliquée, Toulon, France). An air gun is used to drive a free piston which displaces a body of water at such speed that when the piston is arrested a vacuum cavity forms in the water which then implodes. It is the implosion of the vacuum cavity which generates the seismic pulse and as no gas bubble is involved there is no bubble pulse to contend with. The principle is illustrated in figure 5/2. An implosion device based on a different principle is the Vaporchoc seismic source used by CGG (Compagnie Générale de Géophysique). A steam generator on board ship charges a steam tank in the water; an electrically operated trigger valve controls the shooting cycle and at each shot a mass of steam is injected into the sea to form a steam bubble; on closure of the valve the steam condenses and an implosion occurs (see figure 5/2). Both water gun and Vaporchoc depend on implosion of a cavity in the water. A device which aims to produce a seismic pulse by implosion without cavitation is the Flexichoc developed by the Institut Francais du Pétrole and marketed by Geomécanique. This system has the advantage of being a 'safe' system to operate as no dangerous fluids, very high pressure gases or high voltages are used. The principle of operation is schematically shown in figure 5/2. The seismic impulse is caused by the rapid implosion of two circular metal plates: at the beginning of a firing cycle the plates are in the closed position (1) and compressed air is supplied at about 30-70psi to separate them against hydrostatic pressure until the mechanism locks in the open position (2); the device is then primed by extraction of air to form a partial vacuum in the cavity; firing is by an electromechanical trigger mechanism which unlocks the plate separators and allows the plates to rapidly strike towards each other. At completion of the strike, the kinetic energy of the enclosing water is converted into acoustic energy by what is sometimes termed the water-hammer effect.

The seismic sources so far described are all mechanical devices. The sparker source described in the previous chapter has limited application in multi-channel exploration; large sparker units up to 100 kilojoule energy are sometimes used in exploration work for targets a few hundreds of metres deep, but more usually the sparker is only used in exploration of shallow target zones.

The other main group of seismic sources used in multi-channel work are of the chemical explosives type. As noted earlier, large explosive charges are seldom used now. A simple and effective way of using a small charge is to deploy lengths of explosive cord each terminated with a detonator which hooks onto a firing cable in the water. The system, termed Aquaseis, is marketed by ICI. Solid chemical explosives are also used in the Flexotir (Institut Francais du Pétrole) method; about 60g of explosive charge is injected by water pressure into a shooting chamber, a perforated steel sphere, where it is fired electrically. The sphere breaks up the gas bubble thus attenuating its effect on the seismic pulse. A gas mixture can also be used as an explosive. Two widely used systems use a mixture of propane and oxygen as fuel, these are Dinoseis (Sinclair Research Laboratories) and Aquapulse (Western Geophysical). In the Dinoseis exploder, two telescoping chambers are driven apart against a compressed air spring by an explosion of the gas mixture detonated by an electrical spark. In the Aquapulse device the explosion takes place inside a heavy elastic sleeve which balloons on ignition of the gases with the gases venting out of the sleeve into the atmosphere as it contracts.

Comparison of seismic sources of the types briefly described above is usually accomplished by analysis and comparison of their seismic signatures. The signature of a marine source is a curve representing a measurement of pressure variation in the sea against time, preferably recorded at a distance away from the source appreciably larger than the source depth. Such a signature is called a far-field signature. The measuring systems should be such as not to distort the signal and should be recorded in a wide frequency band. The main value of a measurement of source signature is that it gives information on the efficiency of the source, on its amplitude spectrum, and on the relative size of secondary (bubble) pulses. In figure 5/3 signatures are shown for a number of sources to indicate the wide variation of signal characters used in exploration work. The single air gun signature shows a series of bubble pulses and the effect of using a wave-shaping kit is well shown in the signature alongside it. Similarly the type of pulse which results from use of an array of twenty-four guns of various sizes is indicated in the third signature. This signature is very similar to those of the second row, whereas that of a single Flexotir gun is more complex. One of the principal objectives of seismic processing is to compensate for the 'imperfections' of the outgoing seismic signal. For comparison, the complex signature (but of much shorter duration) of a 1 kilojoule sparker is shown.

5.3.2 Hydrophone streamers

Hydrophone streamers used in multi-channel seismic surveys are essentially similar to those used for single channel work (see previous chapter) except that each streamer is made up of a number of equally spaced active sections as shown in figure 5/1. Design features are incorporated to maintain a very long streamer (2-5km) at constant depth in the sea, usually about 10m, and to give maximum signal to noise ratio in the frequency band 10-100Hz. Because signals from a large number of channels are being transmitted through the cable, the design must aim to minimise cross-talk between channels. Modern streamers contain up to 120 data channels.

To maintain constant depth along the length of a neutrally buoyant streamer up to 5km long, a number of depth controllers must be fitted. Each controller is clamped onto the streamer and constant depth is maintained by the action of fins servo-linked to a depth sensor. Separate depth sensors in the streamer are used to monitor cable depth and sonic signals can be transmitted from the survey ship to alter the streamer depth by remote control.

Figure 5/3   Signatures of a range of seismic sources.

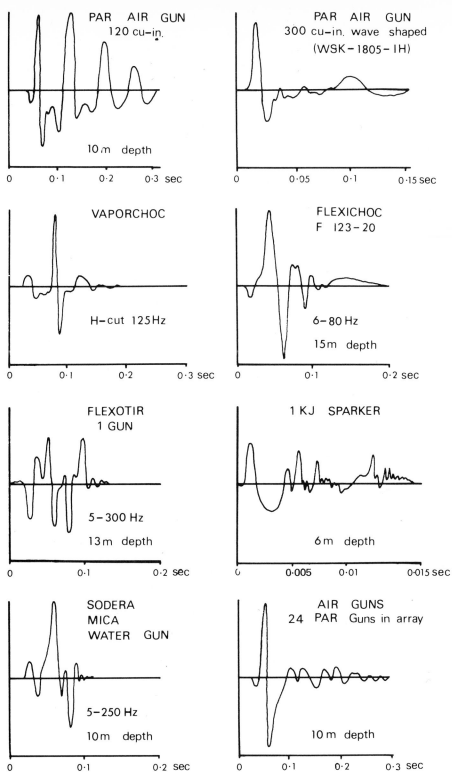

PAR AIR GUN
120 cu-in.

10 m depth

0   0·1   0·2   0·3 sec

PAR AIR GUN
300 cu-in. wave shaped
(WSK-1805-1H)

0   0·05   0·1   0·15 sec

VAPORCHOC

H-cut 125 Hz

0   0·1   0·2   0·3 sec

FLEXICHOC
F 123-20

6-80 Hz

15m depth

0   0·1   0·2 sec

FLEXOTIR
1 GUN

5-300 Hz

13m depth

0   0·1   0·2 sec

1 KJ SPARKER

6m depth

0   0·005   0·01   0·015 sec

SODERA
MICA
WATER GUN

5-250 Hz

10m depth

0   0·1   0·2 sec

AIR GUNS
24 PAR Guns in array

10 m depth

0   0·1   0·2   0·3 sec

93

5.3.3 Recording equipment

The equipment used to gather electrical signals at the output end of a seismic streamer and record these on to digital magnetic tape in a standard format ready for computer processing has a very complex task to perform. It is this recording system which controls the whole seismic process on board ship and the printing onto computer tape of information required for all future processing work. In typical continental shelf surveys, after each shot a record of 5-7 seconds duration is recorded, then the system logs parameters required for the processing and sets up the operational units ready for the next shot. Each seismic record commences just before the shot instant which is itself recorded. Then signals from the hydrophone streamer are accepted, preamplified and filtered, then multiplexed in groups (possibly of 12) before amplification by a gain ranging amplifier. The multiplexed signals are then passed through an analogue to digital (A/D) converter to convert the analogue electrical signals into digital form. The digital signals are then sorted into a standard format and recorded on digital magnetic tape. To monitor the quality of the data

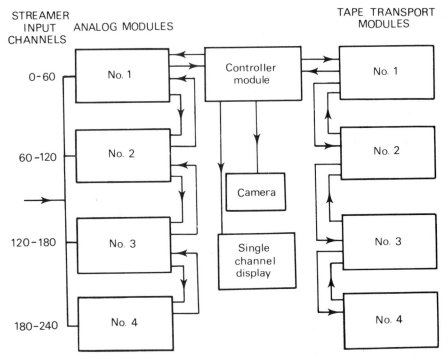

Figure 5/4   Schematic block diagram of the principal components of a Texas Instruments DFS V seismic recording system.

and allow an on-ship evaluation of the geophysical results, equipment is usually included which automatically converts the recorded digital record into an analogue signal which can be displayed immediately as an unprocessed seismogram. This 'monitor' seismogram will probably look similar to the record produced as the final output record of a single channel seismic profiling system as discussed in the previous chapter. Because of the need for flexibility between differing applications, most commercial systems are of modular design so that equipment used at one time for twelve or twenty-four channel work can be easily expanded for ninety-six channel work, or in some very recent designs for work requiring up to two hundred and forty channels. In figure 5/4, the way in which a modern recording system, the Texas Instruments DFS V System, can be expanded to handle increased input and output requirements is illustrated.

Let us now study in a little more detail the main functions of each of the modules, and assume that a streamer of sixty channels or less is in use. Figure 5/5 shows an oscillograph record of the electrical signals gathered from a twenty-four channel streamer. Signal levels output from a streamer can be from a few tens to a few hundred millivolts. The streamer is connected directly to the analogue module in which the signal from each individual channel receives analogue amplification and is passed through high and low-cut filters with the option of notch filters to remove mains pick-up. The group of signals is then multiplexed before analogue to digital conversion and transmission out to the controller module. The analogue to digital conversion process is illustrated in figure 5/6. The electrical signal is sampled at a fixed interval (usually 1, 2 or 4ms) in this case every 4ms, and the amplitude of the signal, positive or negative, is converted into an integer in the range ± 8000 which is transmitted to the next stage of the system in binary form. Thus the varying electrical voltage signal is transformed into a series of numbers n_1, n_2, n_3 ... n_{25} etc. It is a feature of this process that the data will be spuriously modified during processing if the sample interval is such that fewer than two samples are taken per cycle of the highest frequency represented in the signal. Thus it is necessary to filter the signal with a hi-cut filter before sampling; this filter is called an anti-alias filter. For 4ms sampling, frequencies higher than 125Hz must be removed, for 2ms higher than 250Hz and for 1ms higher than 500Hz. The analogue module also has facilities for testing for electrical continuity and leakage in each of the streamer channels. In the controller module, system timing and control is provided, signals undergo digital amplification, then further conditioning before transmission to the tape transport module, now re-ordered into an industry standard format. The controller module also provides

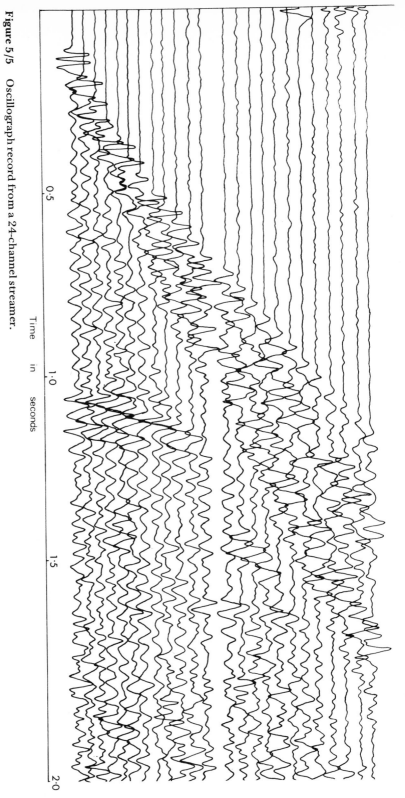

Figure 5/5 Oscillograph record from a 24-channel streamer.

Time in seconds

0·5 1·0 1·5 2·0

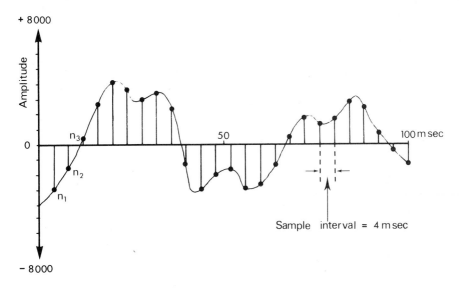

Figure 5/6   Digital sampling of a seismic signal.

facilities for receiving read-after-write data from the tape-transport and converting this into analogue signals for display on an oscillograph, seismic camera or single-channel recorder. Thus it is possible to check and continuously monitor the quality of data being recorded at the final output stage, the digital magnetic tape. The tape transport module is a tape drive with read-after-write facility and interface circuitry which allows automatic rewind and switch-over when two modules are in use as well as circuits which interpret commands from the controller module. The system described above is of very recent design, and is compact and transportable, see figure 5/7. At the time of writing, the majority of systems in use in offshore exploration are physically bigger, with a larger number of modules, but their function and mode of operation are very similar (eg, the Texas Instruments DFS IV, the SN 338 B system made by Sercel, Nantes, France; the GS-2000 system made by Geo-space, Houston, Texas). Although modern systems can record in excess of 100 channels, the costs involved in shooting and processing such high multiplicity data limit the general use of this facility. In the North Sea, by far the majority of data recorded between 1970 and 1976 is in 12, 24, or 48-fold format.

One notable feature of digital seismic work is that it produces very large quantities of digital magnetic tape. Systems have become available in recent years which use specialised tape recording techniques to give substantial savings in tape use (eg, GUS 8DDR system manufactured by GUS Manufacturing Inc; LRS Manpaq system manufactured by Litton

Figure 5/7 Texas Instruments DFS V seismic recording system.
(Photo: Techmation.)
Figure 5/8 LRS Manpaq seismic recording system fitted in Landrover.
(Photo: Western Geophysical.)

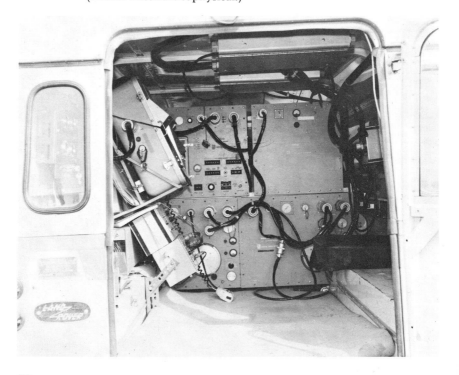

Resources Systems (figure 5/8), both of Houston, Texas). However, the value of such savings must be balanced against the inconvenience, cost and time involved in transcription of the specialised tape on to standard format tape before seismic processing can take place.

5.4 Processing of seismic reflection data

The object of seismic processing in exploration work is to produce a visual display showing a highly resolved reflection pattern at every CDP line (vertical locus of CDP points) assembled into a seismic section so that, as near as possible, interpretation may result in provision of an unambiguous geological section of the earth's crust down to or beyond economic basement. The processes involved in transforming the digital equivalent of the seismic signals shown in figure 5/5 into the seismic section shown in figure 5/9 are not simple and involve a wide range of advanced mathematical techniques. Here, we will not delve into the mathematical basis of many of these processes. A typical processing scheme can be thought of as consisting of a number of stages:

Stage 1	Conversion of field tape data into a data set organised ready for subsequent processing.
Stage 2	Analysis of the data with information print-out to aid selection of optimum processing parameters in stage 3.
Stage 3	Main processing phase in which primary reflection signals are enhanced while multiple and reverberation signals are repressed and noise is minimised.
Stage 4	Digital to analogue conversion and print-out of processed section by graphic plotter or camera.

In stage 1 the main processes include demultiplexing (putting together again complete traces from the data scanned on the field tape) and recovery of the true amplitude relationships (TAR) throughout each trace. The TAR process not only corrects for the effects of time variable gain in the ship recording system but also compensates for attenuation of the seismic signal through spherical divergence and absorption in sea and rock. Finally, CDP traces are gathered into adjacent sets ready for the next stage of processing.

In stage 2, an analysis is made of the data to determine variation of seismic velocity throughout the reflection time interval to be displayed on the final section. If we refer to figure 5/1 again it can be seen that at CDP 1 a number of seismic records, S_1 trace 1, S_2 trace 2, ... S_{48} trace 48 all contain reflections from the same reflection point, but that the distance travelled by the sound along the path S_1, CDP 1 and streamer

Shot point number

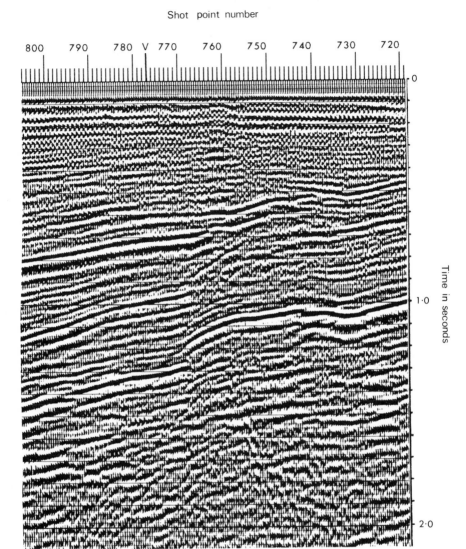

Figure 5/9   Part of a seismic section obtained during an IGS survey of the Moray Firth, off the NE of Scotland. Survey by Seiscom Ltd. (IGS record: Seiscom survey.)

section number 1 is much less than (and therefore takes less time than) the distance travelled along the path S_{48}, CDP 1 and streamer section number 48. If all the traces from a CDP stack are plotted side by side the reflections from a particular horizon form a hyperbola, the curvature of which is a measure of the average velocity of the media traversed by the seismic ray. For the simple case of horizontal rock horizons, the equation for this hyperbola is given by:

$$T^2 = T_0^2 + X^2/V^2$$

Where T is the measured two-way reflection time, X is the horizontal distance between shot and streamer section, and V is the mean velocity, T_0 being the reflection time for $X = O$.

Velocity analysis involves statistical comparison of the set of seismic traces using a range of values of V for the known values of X. These values of V, derived statistically by root-mean-square methods are known as RMS velocities and in practice approximate closely to mean velocities. The velocity determination method then displays results (there are a great variety of ways in which this may be done) so that for a range of NMO (Normal Move Out) velocity values, the coherency of added seismic signals at various levels in the section is indicated, see figure 5/10. It is thus possible to calculate the seismic velocity of any interval in the seismic section as well as decide upon the best set of values of V to use in correction of the seismic traces for normal move out, often referred to as the 'dynamic correction'. It is only when the seismic traces have been corrected for move out $(X = 0)$ that the set of seismic traces can be added together to give a CDP stack. In our example, all forty-eight traces would be added to give a 48-fold CDP stack. The main effect of performing such a stack is that real reflection signals add together coherently to give an amplified signal, whereas seismic noise, electrical noise, and signals formed by multiple reflections do not add and the result is that the real reflection events are very much enhanced. Velocity analyses are usually carried out at intervals of a few kilometres along every seismic line of any survey project. Other tests made in stage 2 include time-variable filter tests, with print-outs, to show the effect on test sections of using various filter settings, as well as deconvolution tests. Filter tests simply aim to apply time varying filters to the processing to give clearest reflection quality at all levels in the section. In the seismic section, figure 5/9, for example, a higher band-pass filter is used in the upper part of the section than in the lower part. The deconvolution process is mathematically rather complex and is one of the most costly, in terms of computer time, of the processes applied to seismic data. The first step is to define a set of numbers, an operator, which when multiplied with the set of numbers which constitutes the seismic trace, or CDP stack, will act as an inverse filter and nullify the effects of signals generated by multiple reflections in the sea water layer as well as reverberations and ringing in the rock layers, see figure 4/17, page 75. Deconvolution tests usually amount to testing different operator lengths and designs and then processing test data to final print-out stage

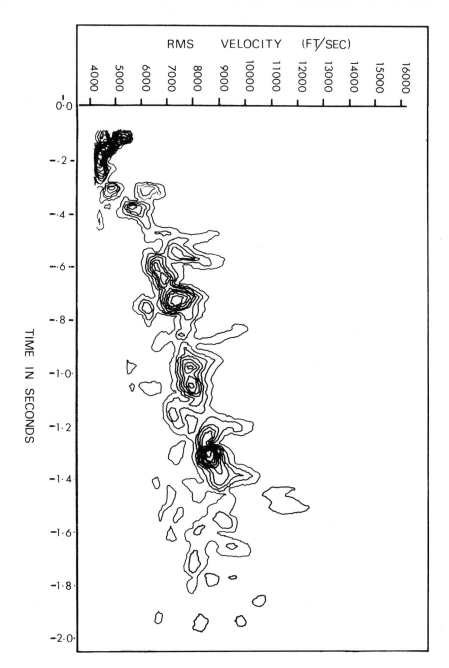

Figure 5/10 Velocity analysis showing coherence of stack at shot-point 830 in figure 5/8.
(IGS record: Seiscom survey.)

from various locations in the survey area. In stage 3 the processing package, which has now been designed, is applied to the results of an entire survey. A typical processing sequence is listed below as applied to data used to produce the section in figure 5/9. In figure 5/10, a velocity analysis is shown; contours indicate coherence of stack over a range of velocity values throughout the two-way reflection time range of the section (figure 5/9) at shot point number 830.

Now that processing is complete, it remains to convert the digital data into analogue form and use the resultant analogue signal to drive a plotting device. Instead of presenting the data as a set of squiggle traces, it is current practice to use a variable area display as in figure 5/9. This type of display very successfully enhances the trace to trace correlation of seismic events.

The above description is a grossly simplified outline of the main elements of seismic processing. Many millions of words are written and published each year on this subject and many thousands of man-hours are spent each year improving and developing processing techniques. We shall now discuss how displayed sections are analysed and interpreted in terms of geological structure.

5.5 Analysis of seismic sections

Part of a fully-processed seismic section is shown in figure 5/9. This example is taken from a survey of the Moray Firth area in the northern North Sea conducted by Seiscom Ltd for the Institute of Geological Sciences in 1972. The project was a reconnaissance survey with line spacing on average 10-15km, total area of approximately 2500km^2; the section in figure 5/9 showing only 4.3km of the total of 600km surveyed. Sections were all processed and displayed to 4s two-way time. In figure 5/9 only the uppermost 2.3s is shown as identifiable reflectors do not occur below this depth at the locality of this record. This section was obtained using an array of eight gas guns as source, two hydrophone streamers providing forty-eight hydrophone groups with a 50m separation between groups, data being recorded using a 48-channel binary-gain digital receiver at a 4ms sampling rate down to 6s two-way travel time. The principal operations in the processing sequence were amplitude recovery, 2-fold sum, deconvolution before stack, static corrections for source and geophone depth, 24-fold CDP stack, deconvolution after stack, time variant frequency filtering, amplitude equalisation and display. Stratigraphic identification of the labelled horizons (figure 5/10) is mainly achieved by study of the results of shallow drilling and seabed sampling at localities where these are exposed at or approach closely to the seabed. In other areas of the North Sea, such identifications are usually based on correlations between seismic sections and data from exploration boreholes within the area of study. Correct stratigraphic identification of the principal reflecting horizons in an area is of course essential both to geological interpretation of the data and for any assessment of the hydrocarbon prospectivity of an undrilled structure. In the early stages of exploration of an area it is unlikely that any borehole data will be available, in which case stratigraphic identification is based on a process of inference and intelligent guesswork using all available circumstantial evidence, such as the results of other geophysical surveys, knowledge of regional geology, and comparison between seismic data and stratigraphy in other more fully investigated areas of similar structure and geological environment. When an exploration well is drilled, it is usual to conduct a geophysical survey in the hole to determine an exact correlation between two-way reflection time and the geological section. This can be done by lowering a geophone into the borehole and measuring the single-way time to a source which is fired at intervals in the sea close to the drilling platform. The shooting interval is fixed with respect to geophone position in the borehole so that a range of travel time measurements are made throughout the depth span of the hole. Alternatively a sonic well-logging tool can be used. The tool is lowered to the bottom of the hole and as it is drawn upwards continuous measurements of seismic velocity are made in the small volume of rocks which form the wall of the borehole. This method has the advantage of giving a sensitive indication of changes in acoustic impedence in the borehole section; thus it indicates interfaces at which strong reflections are likely to occur. Conversion to two-way time is computed by downward integration of the interval velocities.

Having discussed the means of identification of reflectors let us return now to our treatment and interpretation of seismic sections. Using a complete survey, the important horizons are marked up (picked) on the sections and within a network of lines internal consistency must be obtained by checking line inter-sections and picking horizons round a series of closed loops. Faults and other geological boundaries are also identified and marked on the sections. In figure 5/12 an interpreted profile is shown for the whole 60km of seismic section across the Moray Firth which includes the short section shown in figure 5/11. Maps are then prepared for each selected horizon showing two-way reflection time contours and the major structural elements. These are called isochron maps. Figure 5/13 shows an example of such a map for

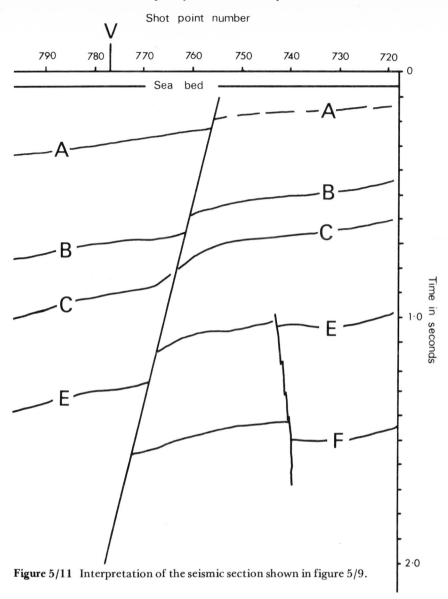

Figure 5/11 Interpretation of the seismic section shown in figure 5/9.

LINE 6

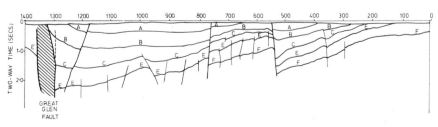

Figure 5/12 Interpreted profile from a survey of the Moray Firth.
(From: Chesher, J A and Bacon, M, 1975. 'A deep seismic survey in the Moray Firth'. *Rep. Inst. Geol. Sci.* No. 75/11.)

the base Jurassic horizon in the Moray Firth. In hydrocarbon exploration a first objective is to identify on such maps areas of positive closure, that is areas which have the topographic equivalent of hill tops and ridges. If such a closure occurs at sufficient depth in a hydrocarbon province, and is likely to contain potential reservoir rocks and be

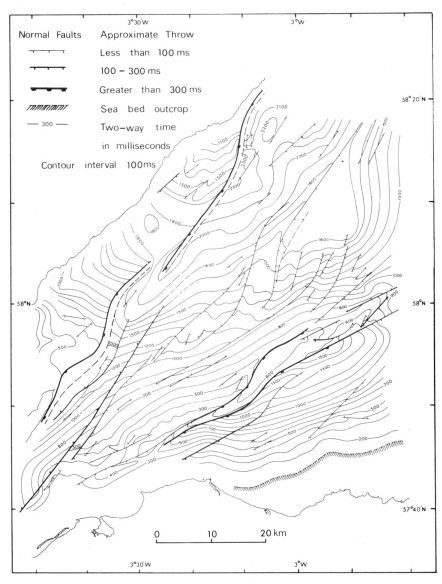

Figure 5/13 Isochron map of the base Jurassic in the Moray Firth.
(From: Chesher and Bacon, 1975. 'A deep seismic survey in the Moray Firth'.)

overlain by an impervious cap rock, then the structure would be identified as worthy of future investigation which might include the drilling of an exploration well. Before drilling, a much more detailed survey than that undertaken by IGS in the Moray Firth would be conducted. Usually a 1-2km grid of lines would be surveyed over the whole prospect with even greater detail near any selected drill site. Such a survey would also be used to make a first estimate of the size of reservoir if oil was discovered in the exploration well, or wells.

So far, we have discussed only isochron maps. If sufficient velocity data are available it is possible to convert isochron maps into depth and/or isopach maps; maps contoured in metres instead of seconds two-way time. However, it must be realised that depth contour values will be subject to quite large uncertainties at localities distant from borehole control due to the commonly occurring lateral variations in seismic velocity. For this reason, depth contour maps have limited and rather specialised use in hydrocarbon exploration. On the other hand, maps showing thickness variation of important reservoir formations can be derived from an analysis of seismic sections and these have an important function in evaluating hydrocarbon prospectivity.

In our brief introduction to seismic processing it was not possible to describe the wide range of specialised techniques used to aid interpretation and enhance quality, usually as a follow-up to the original processing, and usually only over selected parts of profiles where critical evidence is needed by geophysicist, interpreter, geologist or reservoir engineer. Two such techniques, which are very widely-used, deserve mention and serve to illustrate the way in which specialised processing can solve particular problems. The first is concerned with identification of very high amplitude reflections commonly associated with the presence of gas in rocks and sediments. The interface between a water-saturated rock and a gasified rock has an abnormally high reflection coefficient. As many oil pools are topped with gas, a means of identifying high amplitude reflections associated with potential trap structure is a valuable aid to discovery of both oil and gas reservoirs. These high amplitude reflections are called 'bright spots' and special processing packages have been designed to detect and display such amplitude anomalies. The second specialised process is concerned with the problem of acoustic diffraction patterns. In an area where many faults affect the seismic section, diffraction echoes can greatly confuse the seismic section as originally displayed after standard processing. Special processing procedures called migration processing can be used which remove these echoes and thus greatly clarify the true geological relationships between reflecting horizons which have been displaced by faults. In a further refinement of the

F

migration process sections can be depth-corrected automatically and displayed in this form.

5.6 Applications

Multi-channel seismic exploration is costly to perform both in terms of data acquisition and processing. The principal application is in exploration for hydrocarbons where target structures range in depth from approximately 1-5km. The method is capable of providing detailed information on structure, even at the deepest levels of major sedimentary basins, though resolution in general decreases with depth. In conjunction with the results of exploration drilling and geophysical tests in boreholes the data can be used not only to prepare structural maps but also to map facies variations in rock units across large areas. Even when drilling control is absent, rock types can be predicted by a study of seismic character, internal structure and seismic velocities. All such information is important in evaluation of hydrocarbon prospectivity and the selection of sites for exploration drilling.

On a more limited scale, the method is employed by government and research organisations in support of regional structural mapping programmes. The method has also been used to study offshore mining problems, and a more specialised application that has developed in recent years involves using high resolution systems to study near surface structure. For these applications short hydrophone streamers are used and a short sampling interval, usually 1ms, is essential. Sections are produced which are broadly equivalent to continuous seismic profiling sections but with greatly improved data quality. Often both amplitude equalised and true amplitude sections are produced. Results are used in engineering site surveys where drilling rigs or production platforms are to be located. The objective is to identify drilling and foundation engineering hazards such as high level gas accumulations or buried channels or any other abrupt lateral variation in near surface rock and sediment distributions.

5.7 The seismic refraction method

Refraction experiments are carried out for a number of purposes, including those of:

    1)   obtaining data on layering beyond the depth range of the

seismic reflection method,
2) studying deep structure at lower cost (but in less detail) than by the reflection method,
3) obtaining velocity data in the absence of well velocity surveys.

Refraction of seismic waves is governed by the same physical law as the refraction of light waves in optics, Snell's Law. This law is illustrated in figure 5/14 and is stated as $V_1/V_2 = \sin a/\sin \beta$ where V_1 and V_2 are the seismic velocities either side of a boundary, and a and β are the angles of incidence and emergence across the boundary. For the case $\beta = 90°$, a_c is said to be the critical angle of incidence; thus $a_c = \sin^{-1} V_1/V_2$. In a multilayer earth, a distant receiver will detect refracted waves generated by vibration of the refracting interface. The method utilises this effect and a receiver is stationed at a fixed point while a source is fired at intervals along a profile commencing close to the receiver and continuing until either the receiver signals are out of range, or, sufficient data has been collected to satisfy the experimental objectives. For horizontal layering, a time-distance graph of the first arriving seismic signal at the receiver has the characteristics shown in figure 5/15. Thus it is possible to derive information from a time-distance graph both on layer thickness and velocity. For example,

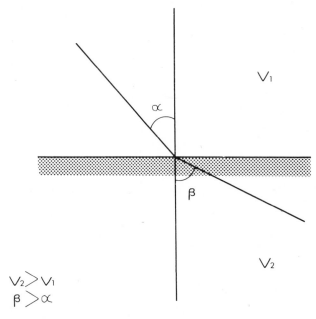

Figure 5/14 Refraction of seismic waves according to Snell's Law.

a three layer case is shown in figure 5/15 where $V_1 > V_2 > V_3$ and the layers have thicknesses d_1, d_2, d_3. The graph will have straight line segments of slope $1/V_1$, $1/V_2$ and $1/V_3$. If the intercepts of these segments through the time axis are plotted, then the first segment passes through $t = 0$, the second segment through t_1 such that,

$$d_1 = \frac{V_1 t_1}{2 \cos \alpha_1}$$

and the third segment through t_2 such that,

$$d_2 = \frac{V_2[t_2 - t_1(1 + \cos 2\alpha_3)/(2\cos\alpha_3 . \cos\alpha_1)]}{2\cos\alpha_2}$$

where,

$$\alpha_1 = \sin^{-1}(V_1/V_2), \ \alpha_2 = \sin^{-1}(V_2/V_3) \text{ and } \alpha_3 = \sin^{-1}(V_1/V_3).$$

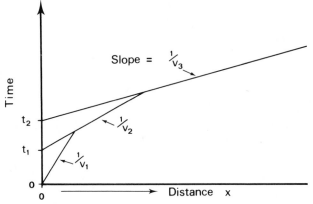

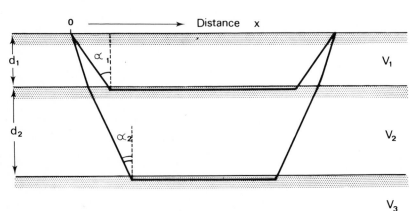

Figure 5/15 Refraction of seismic waves in a three-layer earth.

These relationships can be adjusted and extended for dipping interfaces and for more than three layers but with increasing complexity of the mathematical relationships. To correct for the effects of dip, refraction experiments are usually reversed, that is, the experiment is performed firstly as shown in figure 5/16 with H_1 being the hydrophone site and S_1, S_2 etc being the shot points, then, at the extreme range of the shooting programme a second hydrophone H_2 is sited and the shooting is continued along the profile in reverse direction from H_2 back to H_1. Lack of symmetry between the resulting time-distance curves gives an indication of the presence of dipping horizons and by analysis, dip values can be calculated.

In practice, the most common method of performing refraction experiments at sea is by using sonobuoys as receiver-transmitters. The source is usually a conventional seismic source, though explosives are often used in long range experiments. The sonobuoy is a free-floating or anchored buoy from which a hydrophone is dangled in the sea on an elastic suspension, the buoy containing an amplifier and radio transmitter to transmit seismic signals to a radio receiver on board ship. The shipboard receiver displays the acoustic signal on a multi-channel recording oscillograph. On the oscillograph the shot instant is recorded on one channel, a timing signal on another, and the received signal from the sonobuoy on another. The oscillograph records show, as well as the first arriving seismic signal, a later strong high frequency signal which is the acoustic impulse which has travelled directly through the sea water from source to hydrophone. Two time measurements are made: shot instant to first arrival and shot instant to direct water-wave arrival. The second time divided by the velocity of sound through sea water gives the distance between shot point and sonobuoy.

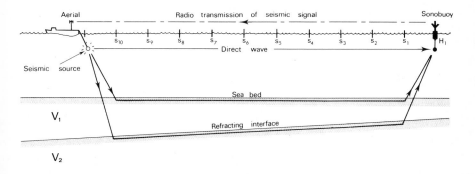

Figure 5/16 Layout of a seismic refraction experiment at sea.

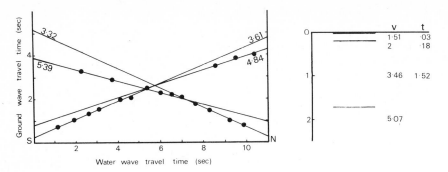

Figure 5/17 Results of a seismic refraction experiment in the Irish Sea.
(From: Bacon, M and McQuillin, R, 1972. 'Refraction seismic surveys in the North Irish Sea'. *J. Geol. Soc. London*, vol. 128, pt. 6, pp. 613–21.)

In figure 5/17 an example is shown of the results of a refraction experiment carried out in the Irish Sea to investigate the causes of a number of gravity lows. In the example shown, the results indicate a layer of approximately 2km thickness of velocity 3.46km/s overlying rocks of velocity 5.07km/s. These results are interpreted as a 2km thick layer of Permo-Triassic sedimentary rocks overlying a Palaeozoic basement. This example is typical of the use of the seismic reflection method as it is most widely applied to investigation of continental shelf geology.

6 Magnetic Methods

6.1 Introduction

Unlike the acoustic methods discussed in the previous three chapters, magnetic and gravity methods do not require the generation or transmission of any signal as part of the operational procedure. These methods depend only on making very precise measurements of small variations in the earth's naturally-occurring magnetic and gravitational potential fields. By comparing observed field strength variations with whole-earth theoretical reference fields, local anomalous variations are detected which can be interpreted in terms of local geological structure. Thus, the magnetic method depends on measuring geographical variations in the geomagnetic field and relating the anomalies thus detected to geological structure involving rocks of differing intensity of magnetisation.

The magnetic method has been in use in prospecting for ore-bodies for over a hundred years. Early instruments measured variations in either the horizontal or vertical components of the earth's magnetic field. In more recent years instruments have been developed which measure the total magnetic field to a very high degree of precision, and this chapter will deal only with the application of such instruments to offshore exploration. Applications of the method are wide-ranging; it can be used to study very large-scale geological structure, such as variations in depth of burial of crystalline basement beneath a few kilometres of sediment; or, it can be used to study very small scale structures such as the course of an igneous dyke, only a few metres thick, and buried beneath a few tens of metres of cover. A magnetometer can be used also to detect such man-made objects as pipelines buried beneath a shallow cover of sea-bottom sediments.

The main application is however as a relatively inexpensive reconnaissance exploration method, and in this mode it is very widely used. In such studies, surveys are more economically conducted using aircraft rather than ships. Airborne surveys are cheaper than shipborne surveys, and can be better controlled, thus providing more accurate data. The importance of the magnetic survey method is evident from the fact that, for example, the entire land area of the United Kingdom, as well as all adjoining continental shelf and margins, are covered by fairly detailed aeromagnetic surveys. Line coverage is quite dense, 2km spacing with 10km tie-lines over much of the area, but expanding to wider spacing in places. In part these surveys have been commissioned by IGS, in which case the results have been published. In other areas, surveys have been undertaken by exploration companies as either group-surveys or speculative surveys, the results in both cases being available for sale to industry. The present position is that if a company requires reconnaissance magnetic data in the UK offshore area, with few exceptions it would be able to immediately acquire such data without the need for further survey work. This does not mean that no further magnetic surveys are being made around Britain; the method has other more detailed and more specialised applications, and is still widely used as a shipborne method in academic research, in work by government agencies, and in hydrocarbon exploration and associated engineering applications.

6.2 Basic principles

A magnet is a body or mass of iron, or other material, which has the property of attracting, or repelling other bodies of similar material.

The force between magnetic poles, either of attraction or repulsion, is given by the equation:

$$\text{Force} = \frac{A.m_1\ m_2}{r^2},$$ where m_1 and m_2 are the strengths of the respective magnetic poles, r is the distance between the poles and A is a constant which can be chosen to be unity depending on the system of units used. Our first basic principle is then that magnetic fields are subject to the inverse square law; the strength of a magnetic field is inversely proportional to the square of the distance from the source. We shall see in the next chapter that the gravity field is subject to the same law.

Electrical currents which flow deep in the earth, possibly associated with convection currents deep in its interior, are regarded as the most likely cause of earth's main geomagnetic field. This field can be represented approximately as dipolar, or as a hypothetical bar magnet, situated at the earth's centre with magnetic moment pointing towards the South Pole. Unlike the earth's gravity field, the magnetic field displays quite large time-varying changes. Such variations are consistent with the theory that the field originates as a result of a dynamic rather that a static condition of the earth's interior. These internally originating variations are long term, and slowly effect both strength and direction of the magnetic field at a given point on the earth's surface. Other, generally externally originating changes are short-term, such as the diurnal effects which are caused by electrical activity in the ionosphere. A further effect, which can be a great nuisance during magnetic survey work, is linked to sunspot activity and is known as magnetic storm activity; sunspots generate showers of high-energy particles of matter which bombard the earth's atmosphere and cause rapid irregular variations in its magnetic field. Amplitudes are commonly of a few hundred gamma, sometimes thousands of gamma. During such activity it is usually necessary to suspend magnetic survey operations or, in compiling maps, to disregard data collected during such periods.

A gamma is 10^{-5} oersted, the oersted being the cgs-emu system unit. One oersted is the field which would exert a force of 1 dyne on a unit magnetic pole. Because the earth's field is dipolar, not only does the

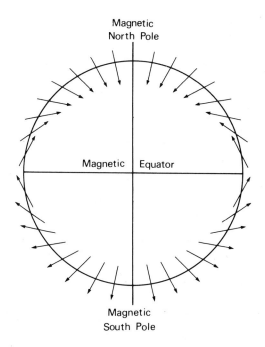

Figure 6/1 Variation in inclination of the earth's magnetic field.

intensity vary across the earth's surface but so does the inclination. At the magnetic equator the direction of pull is horizontal; at a magnetic pole it is vertical; see figure 6/1.

As magnetic survey work depends on the ability to separate local magnetic anomalies associated with geological structure, from regional variations of the earth's magnetic field, it is necessary first to define this regional field. Here, one of the main difficulties is that of accounting for time-varying effects, thereby allowing for comparison and compilation of data observed during surveys which may have been conducted over short periods, but separated by a number of years. Present-day surveys are, with few exceptions, referenced against the International Geomagnetic Reference Field (IGRF) which is defined in terms both of values of the main field and in terms of secular variations. In figure 6/2 the IGRF for the area around the UK is shown for the epoch 1970.0 (ie 1 January 1970).

Earlier surveys around UK were based on a reference field used by IGS in preparation of the Aeromagnetic Map of Great Britain, which is a linear equation in national grid co-ordinates. The IGRF is much more complex and describes the earth's magnetic field in terms of spherical harmonic co-efficients. Tables are available for conversion from one system to the other.

In figure 6/2, field values are quoted in gammas which are the units used in most exploration work.

From equator to poles, intensity varies approximately in the range 0.4-0.6 oersted, ie 40,000-60,000 gamma.

6.3 Survey magnetometers

Although the magnetic survey method depends in most situations only on the need to make relative determinations of magnetic field variation, the instruments most commonly used in such work in fact provide absolute total field determinations. Good quality survey data can be acquired using magnetometers with perhaps less difficulty than is the case using any other geophysical instrument in general use in the exploration of continental shelf areas. Instrumentation is relatively simple, not excessively expensive, and surveys are only weather dependent in so far as aircraft or ships are weather limited.

Early shipborne and airborne surveys were made using flux-gate magnetometers which gave a generally adequate accuracy of 1 gamma but were less simple to operate than modern proton magnetometers. Present day surveys are almost exclusively conducted using proton magnetometers which can give an accuracy of better than 1 gamma. If even greater precision is required, optical pumping magnetometers can be used the accuracy of which can be as high as 0.01 gamma.

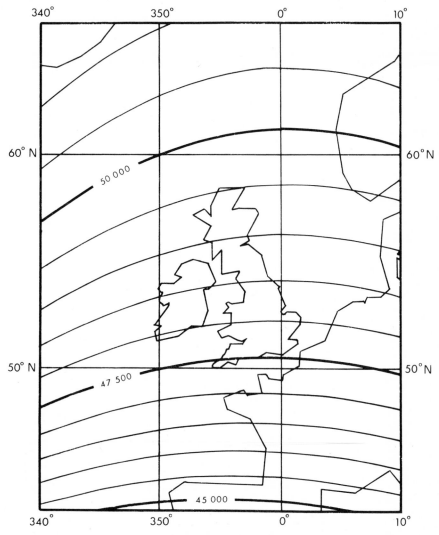

Figure 6/2 IGRF values of total intensity, F, in gamma for the epoch 1970.0. (Adapted from figure 7 in Barraclough, D R and Malin, S R C, 1971. 'Synthesis of International Geomagnetic Reference Field values'. Rep. No. 71/1, *Inst. Geol. Sci.*)

Proton magnetometers use a sensing element which can be simply a bottle of water around which is wound a coil of wire. Operation of the instrument is based on the principle of nuclear precession. A proton is an elementary particle of postive charge and unit atomic mass; an atom of the lightest isotope of hydrogen if separated from its electron is a proton. In the magnetometer, protons exist in the water bottle as hydrogen ions and these, due to the fact that they spin about a magnetic axis, have a tendency to be aligned parallel to the prevailing magnetic field. To make a magnetic measurement, a large current is passed through the coil encircling the bottle to produce a strong magnetic field in a direction substantially different from the earth's field thereby aligning protons in this direction. The current is then cut off abruptly, and the spinning protons revert into an alignment with the earth's field. In so doing, these protons are subject to a precessional oscillation the frequency of which is proportional only to the strength of the prevailing earth's field. This relationship is described by

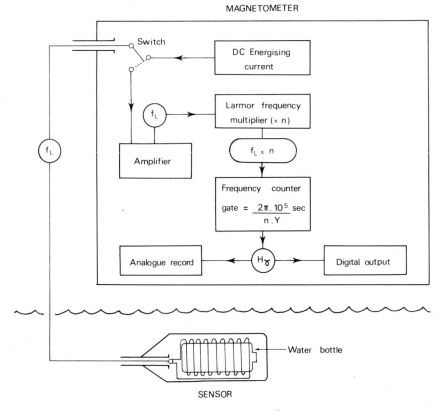

Figure 6/3   Schematic diagram of a marine survey magnetometer.

Larmor's theorum; $H_0 = 2\pi f/y$ where f is the precessional, or Larmor frequency, and y is the gyromagnetic ratio 26,751.3, H_0 being the earth's magnetic field in oersteds. The precessional oscillation of the protons produces a small electrical signal in the coil of wire which can be detected and its frequency measured. Measurements cannot be made continuously, but nevertheless, with a modern marine proton magnetometer it is possible to make discrete measurements with the sensitivity of 0.5 gamma every four seconds along a profile. Such a repetition rate is well within the requirements of most survey specifications.

A schematic diagram showing how a magnetometer functions is shown in figure 6/3. Switching between energising current and measuring circuits is automatically controlled by a clock. The signal of frequency f is first amplified, then, so as to obtain a high accuracy measurement over a short period of time, multiplied through a series of binary multiplying circuits. For the output to be registered in gamma it is necessary to accurately gate the counting period to equal $2\pi . 10^5 /ny$ seconds, ie, 0.7339815 seconds if n = 32.

In figure 6/4 a photograph is shown of a typical marine magnetometer system.

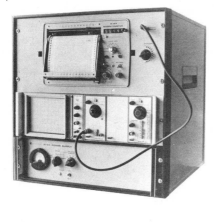

Figure 6/4   A Barringer marine magnetometer.
(Photo: Marine Environmental Services, UK.)

Proton magnetometers are capable of precisions of up to approximately 0.1 gamma. For certain very specialised applications higher precisions are required and a further range of instruments has been designed, again dependent on nuclear phenomena, and in this case based on optical pumping and the Zeeman effect. This class of instruments can give measurements to an accuracy as high as 0.01 gamma. Advantages of the proton magnetometer are low cost, simplicity and ruggedness of the sensor, this being the most vulnerable part of the system, because when towed behind ship, helicopter or aircraft there is always a possibility of damage or even loss.

6.4 Magnetic data reduction

In magnetic surveys where lines are relatively short, up to a few kilometres long, data output from the magnetometer can be immediately plotted as profiles, and, being a series of absolute measurements of the earth's total field, these can be interpreted without application of any corrections. However, if the requirement is that of surveying a large area with a grid of lines to produce a magnetic anomaly map, then the procedure is more complex. During the survey, daily magnetic variations must be monitored, and if magnetic storms occur parts of the survey will need to be repeated. Smaller, regular daily variations are usually compensated for by application of corrections to a wide grid of lines enclosing the area surveyed using a statistical method; a least-squares network adjustment. All infilling lines are then fitted into this framework of adjusted observations. Adjusted values are then compared at points along a profile with IGRF values, and values of differences between observed and IGRF values are plotted on a map. This map is then contoured, usually at 5 or 10 gamma intervals and at a scale of 1:100,000. The result is termed a magnetic anomaly map.

If the data were collected by airborne survey, it is possible to over-fly both sea and land areas during the same project. In such a situation it is usual for marine areas to be surveyed using a radio-navigation system for position fixing, and land areas to be surveyed using a vertical aerial camera, photographs being shot during the flight in synchronisation with time marks on the magnetometer record and a digital recorder. Over the sea, the aeroplane usually maintains a fixed height above sea level; 1000 feet being commonly used around UK; over land areas however, a system of drape-flying is usual, where the aeroplane attempts to fly as closely as possible to a constant height above land surface. The map, figure 6/5, shows data contoured from a survey made in this way.

Marine surveys have the advantage that observations are made at a level nearer to target structures and thereby such surveys can detect smaller features. However, because of the slow relative speed of ships, such surveys are less economic unless combined with other work, or are concerned with detection of very small targets within a limited area. Furthermore, it is not usually possible to attain in marine surveys the high accuracy possible with airborne surveys, due to the difficulty in correcting for diurnal changes of up to 10-20 gamma which can occur between line intersections in the control network. For example, if a survey is based on a 50km control network, flight time between control lines is likely to be about 10-15 minutes (diurnal changes are not likely to exceed 2-3 gamma in non-storm conditions), but sailing time might be five hours (diurnal changes of 10-20 gamma being possible in non-storm conditions). For these reasons, although data from shipborne surveys can be contoured into maps, more usually, the data are presented as profiles which can be directly and accurately correlated with seismic, gravity and bathymetric profiles

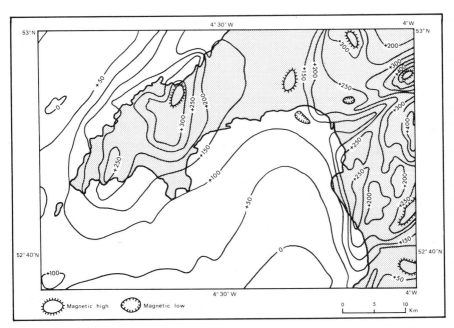

Figure 6/5   Aeromagnetic anomaly map of the Tremadoc Bay area of the southern Irish Sea. Contours in gamma.
(Adapted from IGS aeromagnetic map.)

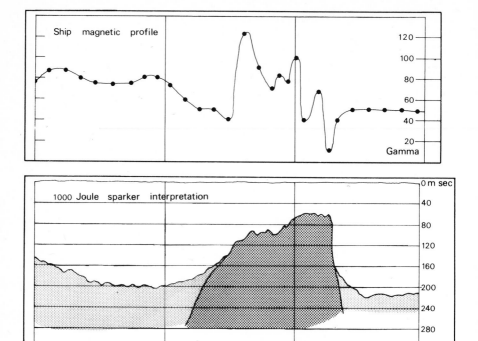

Figure 6/6   Top: Marine magnetic profile across a small seamount. Bottom: interpretation of a sparker profile over the same feature.

simultaneously recorded from the same vessel. In figure 6/6 a marine magnetic profile across a small seamount in the North Channel between Scotland and Ireland is shown correlated with an interpretation of a sparker seismic profile. The seamount is probably a volcanic neck made up of rocks which are more magnetic than the enclosing sedimentary rocks. This magnetic anomaly was first detected by an airborne survey and found to correlate with a topographic feature; the ship magnetic and seismic profile were surveyed as a means of studying the origin of the aeromagnetic anomaly.

6.5 Applications of the magnetic method

In oil and gas exploration, aeromagnetic surveying is the most commonly used reconnaissance method and is employed to define the existence of, and limits to, sedimentary basins in a previously unexplored marine prospect. The method can be used to estimate the thickness of sediments in a basin, as well as to study its structural form and the regional structure of the surrounding economic basement. Results provide a valuable guide in planning relatively expensive seismic reflection surveys. Obviously, no company wishes to expend large sums of money surveying with a seismic ship areas where economic basement is close to the seabed and where seismic reflection records will provide little or no useful information. Commercially, shipborne magnetic surveys are used mainly as an aid to interpretation of seismic surveys being recorded concurrently with the seismic shooting.

Outside the hydrocarbon industry, shipborne magnetic surveys are widely used as an aid to geological mapping particularly by universities and research institutes. In such cases measurements are usually made concurrently with gravity and high resolution seismic measurements.

A more specialised, non-geological application of the method is in research for ferro-magnetic metallic objects at or near the seabed. Magnetometers are used, for example, in searches for buried wrecks and pipelines.

6.6 Geological interpretation of magnetic data

Methods employed in the geological interpretation of magnetic data closely resemble those of gravity interpretation, which will be discussed in the next chapter. In both cases, the methods depend on the ability of the interpreter to define a geologically acceptable model structure, which can be shown by computation to be capable of producing a magnetic or gravity field closely comparable with the field anomaly pattern discovered as a result of exploration. Generally, the model structure is a considerable simplification of that likely to occur naturally, but such is true of any sensible interpretation of geological structure in depth, whether derived from geophysical or purely geological data.

The magnetometer would be a much more powerful exploration tool if it were not for the fact that bodies of rock are usually magnetically anisotropic; even similar rock types exhibit widely varying properties. This variability derives from the fact that the magnetic properties of all common rocks are controlled by the presence of, and history of formation of, ferro-magnetic minerals such as magnetite, which occur only as minor constituents. The size and form of a magnetic anomaly pattern, produced at the earth's surface by a body of magnetic rock at depth, is determined by the size and shape of the rock body, its depth of burial, the strength and direction locally of the earth's magnetic field, the magnetic susceptibility of the rock, k, the permanent intensity of magnetisation of the rock, Mn, and the direction of this permanent magnetisation. A rock body is usually described as having a magnetisation, M (dipole moment per unit volume) which is the sum of two vector quantities, Mn and k.F, where F is the total magnetising force of the earth's field. Mn is a magnetisation which is imprinted in the rock as it forms, either at the time of cooling and mineral formation, such as occurs in igneous and metamorphic rocks, or at the time of deposition of rock grains by orientation of particles, as occurs in sedimentary rocks. The direction of Mn is usually that of the earth's magnetic field at the time of the rock's formation; k.F is the magnetisation which is induced in the rock by the earth's magnetic field and is therefore parallel to it. In the practical task of interpreting magnetic maps or profiles it is usually possible to assume only a very approximate value for Mn and its direction.

Without quoting actual values for k, or probable values of M for different rock types, it is possible to give a general classification of rocks into groups likely to have similar properties, and such a classification is given in table 6.1.

Modern interpretation techniques are based on analysis of large data sets compiled from maps and profiles using digital computers to handle lengthy numerical procedures. A large literature exists explaining numerous variations and refinements available with particular accent in recent years on methods utilising Fourier analysis techniques. By Fourier analysis, a wave form such as a magnetic profile is redefined mathematically in terms of a series of sinusoidal functions which can then be analysed to give information on the nature of the original wave form. The value of this method is that it can be automated and gives an objective interpretation; also, the method can be applied to gravity interpretation, then gravity and magnetic interpretations can be directly compared.

Here, we shall concern ourselves mainly with what might be termed 'first-look' interpretation methods. Such methods allow a simple appreciation of the geological meaning of the results of magnetic surveys. At the end of the book specialist texts are listed which discuss the mathematical basis of more complex modern interpretational methods. Any approach to interpretation is to a large extent controlled by the specific objective of the exploration project. In hydrocarbon

Table 6.1

Magnetic properties of rocks

(M,k) Values	*Rock types*
Very low	Quartzite; most sandstones, limestones, silt-stones, mudstones, unconsolidated glacial and marine deposits; salt and other evaporites; some igneous rocks particularly granites and acid volcanics containing no magnetite.
Low	Some sedimentary rocks which contain small quantities of ferro-magnetic minerals; igneous and volcanic rocks with low magnetite content; some metamorphic rocks.
Moderate	Most volcanic eruptive rocks, tuffs etc; many metamorphic rocks which occur in basement; many acid and intermediate igneous volcanic and intrusive rocks, some basic igneous rocks; some types of iron ore.
High	Many basic igneous rock types, basalts, gabbros etc; some granites and other intrusive rock with high magnetite content; some iron and nickel ore deposits.
Very high	Illmenite and magnetite ore deposits.

exploration the object is usually to calculate depth values to magnetic basement as this often corresponds with economic basement, and to define the positions of faults and other structural elements. In mineral exploration the objective is usually that of locating targets for drilling, and to calculate the size and shape of ore bodies once these have been discovered. In regional geological surveying and research work the objective is usually to provide data which can be used in mapping seabed geology and interpreting structure at depth; to some extent an extension of the objectives of hydrocarbon exploration. In engineering the object is to locate targets which are potentially engineering hazards or obstacles to development of particular engineering projects.

As a first illustration, in figure 6/5, a magnetic anomaly map is shown over Tremadoc Bay on the coast of Wales. The survey was made by IGS to aid structural mapping, but results have been widely used in hydrocarbon exploration. Over land areas enclosing the Bay, large magnetic anomalies relate to igneous or metamorphic rocks at or near land surface, whereas, in the offshore area, the magnetic field is featureless and exhibits only a gentle negative southerly gradient. The indication is that offshore a thick sequence of non-magnetic sedimentary rocks overlie the economic/magnetic basement. Furthermore, the strong linear magnetic gradient which closely follows the Bay's eastern coastline suggests that the sedimentary basin is fault-bounded along this margin. The conclusions arrived at by simple inspection of the map, have been confirmed by subsequent seismic and drilling exploration.

In Tremadoc Bay, the anomaly pattern indicated the presence of a sedimentary basin, but because of the nature of the geological structure, it was not possible to derive a meaningful estimate of the thickness of sediments underlying the Bay. In other areas, where sediments overlie a magnetic basement at varying depths such estimates can be made and major structures mapped. In figure 6/7 a magnetic anomaly pattern over a marine area in the UK continental shelf is shown with, on the same map, a depth to basement and structural interpretation. This interpretation is based on an automated analytical technique. The example is part of a survey undertaken for hydrocarbon exploration. In this case a well-developed anomaly pattern, caused by structure below the base of sedimentary fill, allows a fairly detailed estimation to be made of variations in thickness of this fill.

To gain better understanding of how it is possible to estimate depths of burial of a body of magnetic rock it is necessary to study more closely the relationship between anomaly shape, as seen in a profile across a buried body, and its depth of burial. As examples we shall use firstly two simple models, a buried horizontal cylinder and a buried vertical cylinder. Neither model would appear to closely resemble any common geological structure, yet, as the object is to define depth to top, this lack of resemblance is not very important. Detailed analysis can be used to show that these shapes give good approximations to a wide variety of geological structures in estimating this parameter. Horizontal cylinder models are used to interpret elongate anomalies, vertical cylinder models to interpret anomalies which are more nearly symmetrically circular.

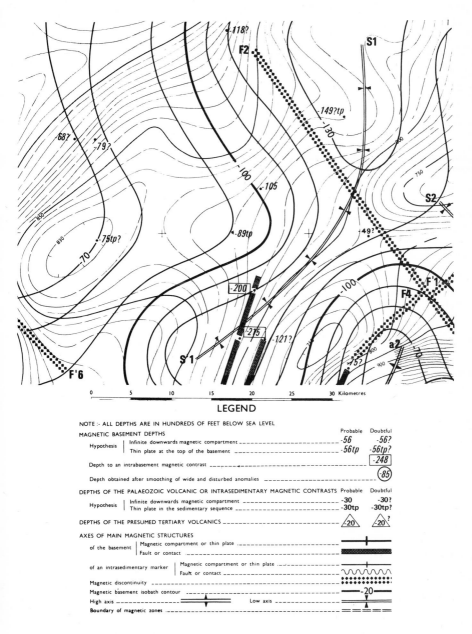

Figure 6/7 Magnetic anomalies and interpretation offshore north-west Scotland. (Part of an aeromagnetic survey by Hunting Geology and Geophysics, UK.)

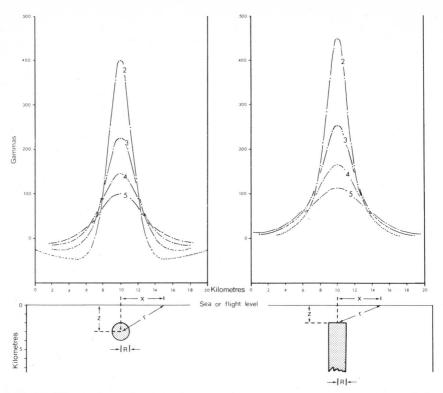

Figure 6/8 Computed magnetic anomalies over simple structures. On the left, over a horizontal cylinder; on the right, over a vertical cylinder. Curves 2—5 represent different depths of burial (Z) from 2—5 km.

In figure 6/8, the anomaly profiles are plotted for both models for a range of depths of burial. The magnetisation of the bodies is, in this case, defined as vertical. The mathematical basis of the computations of the profiles is beyond the scope of this book.* In figure 6/9 a slightly more complicated case is illustrated. A computer has been used to calculate profiles across a horizontal prism, again for a range of depths of burial. The prism has a 2km × 2km section and is 8km long, orientated east to west. Profiles are North-South across the middle of the model. The magnetisation of the body has a northerly dip of 65°.

Even without making any assumption as to the size and shape of the causal structure, a very approximate guide to depth of burial of magnetic basement can be obtained by abstracting a profile along a line normal to the trend of magnetic anomalies and measuring the

* This can be studied in, for example, Grant, F S, and West, G F. *Interpretation theory in applied geophysics*, McGraw-Hill Book Company, New York

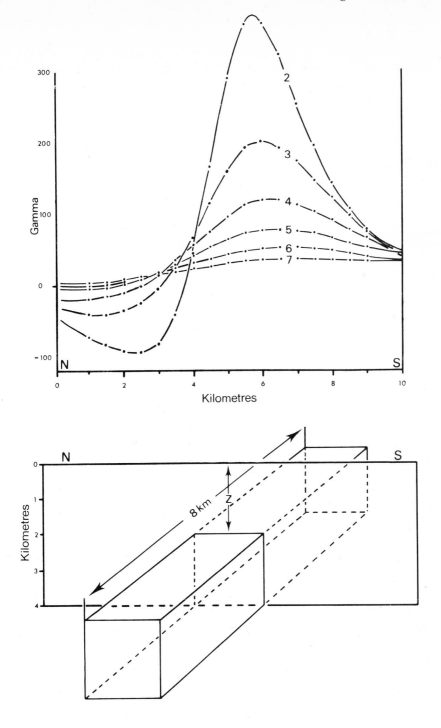

Figure 6/9   Computed magnetic field across a horizontal prism. Curves 2—7 represent different depths of burial (Z) from 2km to 7km.

half-amplitude width of prominent deflections. This gives an immediate indication as to whether magnetic basement is close to the seabed, and can be correlated with, for example, a strong seismic reflector below superficial deposits, or is buried at depths of a few kilometres. It should be noted that it is the 'spikyness' or 'roughness' of the anomaly pattern which is significant, not the size or amplitudes of the magnetic variations.

For precise depth estimates it is essential to undertake more complex interpretation procedure based on either curve or contour pattern matching to a wide range of models, a method which demands considerable practice and experience on the part of the interpreter and is very time consuming; or, automated computer methods which demand a lower input in man hours by experienced personnel, but which still need to be monitored and controlled by an experienced interpreter.

7 Gravity Methods

7.1 Introduction

The gravity method was the first geophysical method to be used in the search for oil and gas, though the original type of instrument to be employed, the torsion balance, is now obsolete in exploration work. First attempts to study geological structure by this method were made in Europe during the early years of the present century, these early experiments being followed by its introduction to the Gulf Coast, USA in the 1920s for use in exploration for buried salt domes. A number of domes were discovered and subsequent drilling led in some cases to the discovery of oil, the first such discovery to be the result of geophysical exploration.

Modern gravity surveys use gravity meters, not torsion balances. A gravity meter is an instrument which measures variations in the earth's gravity field: it does not give an absolute determination of the force of gravity. With a modern land gravity meter it is possible to measure the change in gravitational pull between adjacent sites to an accuracy of one part in 100 million. This type of instrument can be thought of as a very refined spring balance capable of detecting, for example, the change in weight of a very small fixed mass inside the instrument between observations made firstly at ground level, then on top of a table. The mass weighs less at table-top level than at ground level because the measurement is made at a greater distance from the centre of the earth. Though described in simple terms as a spring balance, the measuring element of a gravity meter consists of a complex and finely engineered mechanical system.

Because a ship is always in motion, it is not possible to measure gravity at the sea surface with the same accuracy as can be attained on land using a land gravity meter. If high accuracy is required at sea, it is possible to use a land meter fitted inside a pressurised container which can be lowered to the seabed for observation by remote control. This is a slow and therefore costly survey procedure. In the late 1950s, gravity surveying at sea first became widely applied as an exploration method with the development of instruments which could be operated on survey ships to give continuous profiles of gravity variation along each course surveyed. The survey accuracy thus obtainable, even with present-day refinements, is approximately one tenth that using a land or sea bottom gravity meter.

In geophysical exploration, gravity surveys are made to detect local anomalies in the earth's gravity field, these being caused by geological structures in the upper part of the earth's crust involving rocks of contrasting density. The gravity method can be used to locate only fairly large-scale geological structures; it cannot usually be applied to small-scale problems, such as studies of the engineering geology of the uppermost 100m of rocks and sediments below land surface or seabed. It has its greatest value in the study of the structure of sedimentary basins, the forms of large igneous intrusions, the courses and throws of large faults, and other similar major structures. In such work, it is used both in reconnaissance surveys and as a very valuable backup to seismic reflection surveys.

Densities of common rocks range from approximately 1.8-3.2g/cc, but density contrasts associated with geological structures seldom exceed 0.5g/cc. In considering the applicability of the gravity method to exploration problems, it is useful at the outset to appreciate the size of body necessary to produce a detectable gravity anomaly. Let us assume that the objective is to detect the location of a buried spherical body, 0.5g/cc more dense than the rocks in which it is buried, and hidden beneath a negligible thickness of cover. It should be possible to detect a 0.5 milligal (mgal) anomaly corresponding to a 75m diameter sphere on land or by using a sea bottom gravity meter (see section 7.2 for definition of milligal). Using a ship gravity meter, it is not practicable to investigate anomalies of less than 5 milligal amplitude, and if the survey is made in water of 100m depth, such an anomaly would be caused by a sphere of just over 1km diameter. This low resolving power should be compared with that of high resolution seismic profiling, a method which can detect in good conditions an individual rock boulder of less than 1m diameter buried in clay deposits to depths of a few tens of metres.

7.2 Basic principles

Newton's Law of Gravitation describes how particles of matter attract one another with a force directly proportional to the product of their masses, and inversely proportional to the square of the distance between them, the law being expressed in the equation;

$$F = \frac{Gm_1 m_2}{d^2}$$

where F is the gravitational pull, d is the distance between the masses m_1 and m_2, and G is the gravitational constant which has been experimentally determined as 6.670×10^{-11} $m^3/kg\ s^2$. This equation shows how gravity at the earth's surface is related to its total mass and its shape. If the earth were a perfectly homogeneous non-rotating sphere floating in empty space, then the gravitational pull would be uniform over its entire surface, but the real-world situation is more complex.

Ignoring for the moment local rock structures in the earth's crust, which are the cause of the anomalous gravity variations explored by the gravity method, let us consider the causes of large-scale variations in the earth's gravity field. Firstly, the world is not a sphere but an ellipsoid of revolution with an ellipticity of approximately 1/298.25, thus a site at the Pole is nearer the earth's centre of gravity than one at the Equator. Secondly, the world rotates approximately once every 24 hours; thus any site on the earth's surface is not stationary but is subject to an accelerational force (centripetal force) due to rotation of the earth, and this force, which opposes the gravitational pull, is maximum at the Equator and zero at the Poles. The effects of ellipticity and rotation combine in such a way that at sea level, gravity at the earth's Poles is approximately 0.5% higher than at the Equator. Thus a pound weight weighed by a spring balance on the Equator, will, if measured by the same balance at a Pole, weigh approximately 1.0051lbs, both measurements being made at sea level.

The unit of measurement in exploration work is based on the gal (from Galileo) which is an acceleration of $1cm/s^2$. For practical purposes, this unit is too large and either the milligal (mgal) which equals 1/1000 gal, or the gravity unit (g.u.) which equals 1/10,000 gal are used. In this book the milligal is used throughout, this being still the most common unit used in exploration work. The earth's theoretical or reference field varies from approximately 978,032 to 983,218 mgal at sea level between Equator and Pole. An equation can be written which describes this theoretical variation of gravity with latitude at sea level. Nowadays, in most organisations, maps are prepared using the 1967 International Gravity Formula which is as follows:

$$g\gamma = 978031.8\,(1 + 0.0053024\,\sin^2\gamma\text{-}0.0000058\,\sin^2 2\gamma)\,\text{mgal},$$

where $g\gamma$ is the theoretical gravity value (in milligals) at latitudeγ. Many maps still in existence refer to an earlier formula of 1930;

$$g\gamma = 978049.0\,(1 + 0.0052884\,\sin^2\gamma\text{-}0.0000059\,\sin^2 2\gamma)\,\text{mgal},$$

and care should be taken in compiling or comparing the results of different surveys to ensure that they are not computed against different reference fields.

Thus, the first objective of a gravity survey is to measure gravity variations across the area under study and to subtract from each measured value the theoretical reference field value to give a map of gravity anomalies which can then be related to geological structure. However, this simple approach can only be adopted when all measurements are made on land at sea level using a stationary gravity meter. If the measurements are made, either at different levels throughout the survey, as occurs with sea bottom surveys, or from a moving ship over a varying range of water depth as is the case with surface ship surveys, then a range of corrections must be applied to the observed data. These are discussed later in following sections.

7.3 The sea bottom gravity meter

Gravity surveys on the seabed have limited application, but some instruments are still available and can be used, for example, to survey estuaries where navigation of a large vessel is impracticable, to accurately establish test-range ranges for calibration of shipboard gravity meters, and to make detailed and accurate surveys across known geological structures which are likely to produce anomalies too small to be detected by shipboard surveys.

The most commonly used instrument of this type is the LaCoste and Romberg underwater gravity meter and we will base our discussion of this type of instrument on this particular meter.

The gravity meter shown in figure 7/1, is similar to a land instrument

Figure 7/1 LaCoste and Romberg underwater gravity meter.
(Photo: LaCoste and Romberg Inc, USA). Above: complete system
including onboard control and recording units. Below: gravity meter
being lowered into the sea.

but for underwater use it is sealed in a watertight container, designed for operation at depths of up to a few thousand feet. The instrument incorporates remote control systems, automatic servo-levelling and compensation for seismic disturbance on the seabed caused by wave motion in the sea. The meter is designed to withstand the inevitable rough handling associated with lowering and raising from ship to seabed many times each working day.

The gravity response system attains its high sensitivity to small gravity changes through use of a 'zero-length' spring and this type of instrument is termed unstable.

Unstable gravity meters are designed so that as the mass in the measuring element is displaced from its null position by a gravity change, additional forces act on it tending to increase this displacement; thus the sensitivity of the measuring element is greatly increased. So as to illustrate the design principle of gravity meters and the use of 'zero-length' springs the concepts involved are briefly discussed as applied to the design of the LaCoste and Romberg system.

The 'zero-length' spring is a spring whose effective length, between fixed points of attachment, is zero when external forces acting on the spring are zero. Obviously, in practice, no such miraculous dematerialisation occurs; the zero-length spring only acts as a spring after application of a stress equal to the unstressed length of the spring multiplied by the spring constant. With the addition of greater stress

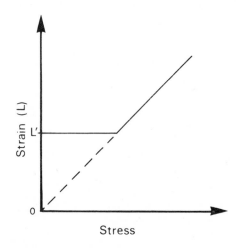

Figure 7/2   Zero-length spring stress-strain graph.

the spring extends following a normal linear stress-strain relationship, see figure 7/2. In reality a zero-length spring looks much like any other spring, the pre-stressing being a feature of the way in which it is manufactured.

The zero-length spring in a LaCoste and Romberg meter is used to support a beam hinged at one end with a mass at the other. The mechanics of the system are shown diagramatically in figure 7/3. This diagram shows an idealised unstable system; the actual design is modified to obtain the required sensitivity for small changes of gravity but engineered to be of rugged construction and to have the necessary stability to allow operation in a range of conditions.

In figure 7/3 the torque at the centre of mass of the beam due to the action of gravity on this mass equals mg.$\overline{AB}$ sinθ, where m is the mass of

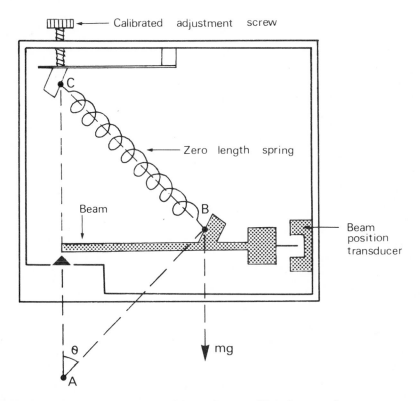

Figure 7/3 Schematic diagram of the LaCoste and Romberg gravity meter. (Courtesy: LaCoste and Romberg.)

the beam and $\overline{AB}$ $\sin\theta$ is the distance between fulcrum and centre of mass of the beam. The system is designed so that θ is close to $45°$, and so that $\overline{CB} = \overline{AB}$. Torque due to the action of gravity on the mass is balanced by the restraining force of the zero-length spring, which = $k.\overline{CB}.(\overline{CB}.\cos\theta)$ where k is the spring constant, and $\overline{CB}$ is both its actual length and its extension. As the system is designed, $\overline{CB} = \overline{AB}$ and at $45°$ $\cos\theta = \sin\theta$, therefore the net torque = $mg.\overline{AB}.\sin\theta - k.\overline{CB}^2.\cos\theta$, or by substitution, torque = $\overline{AB}.\sin\theta.(mg - k.\overline{AB})$. For balanced equilibrium, $mg = k.\overline{AB}$, and for small changes of θ and constant k (ie for small deflections of the beam from the null position) the torque remains approximately zero, the equilibrium is unstable, and the instrument is extremely sensitive to small variations of gravity. We have discussed this matter at some length so that the reader may understand how, in one particular design, the seemingly impossible task of designing a static mechanical system to measure gravity changes to one part in 100,000,000 is actually accomplished. Other types of gravity meter use other equally ingenious design concepts.

The measurement of gravity using the LaCoste and Romberg seabed gravity meter entails the following operations:
1.   The ship is positioned over the required site,
2.   the instrument is lowered to the seabed,
3.   the gravity measuring mechanism is unclamped,
4.   the instrument is operated by a remote control system which initiates automatic levelling and indicates the position of the gravity meter beam,
5.   the calibrated adjustment screw is rotated by remote control until the beam is brought to a null position,
6.   the calibrated adjustment screw dial reading is recorded along with water depth and ship's position in a gravity data book.

Turns of the adjustment screw are calibrated in milligals and the position of the screw is remotely recorded on board ship on a set of counters. The gravity difference between any two sites is thus indicated by the difference in reading of these counters.

7.4 The shipboard gravity meter

There are two main types of surface ship gravity meter, static and dynamic. Although, theoretically, dynamic meters have certain distinct advantages over static types, in practice some designs of the dynamic type have suffered from important operational drawbacks. The result is that the static type of instrument is still in very wide commercial use. The LaCoste and Romberg air-sea gravity meter provides a good example of how far it has been possible to develop the static type of instrument for use on board a ship. We shall discuss this matter in some detail, and indicate how other commercially-available systems compare with it.

To measure gravity aboard a surface ship using a static gravity meter the instrumental system must resolve the acceleration due to gravity, as it acts on a mass inside the meter, from accelerations due to the motion of the ship. This ship motion is complex and different components have different periodicities. Anyone who has experienced rough weather at sea will be fully aware of the way in which g appears to vary as the ship heaves, pitches and rolls. As a first essential, the gravity meter must be kept level and this is accomplished by putting it on a gyro-stabilised platform, see figure 7/4.

This platform is controlled by two high response torque motors, each motor is linked into two servo loops which are fed by output from a gyro and a horizontal accelerometer. The first loop is designed to null the gyro output thus stabilising the platform in space. In the second, the accelerometer output is used to precess the corresponding gyro to maintain its verticality. The precession period must be longer than ocean wave period and is usually chosen as four minutes. The system is specified to maintain a long-term platform verticality better than 1s of arc through an angular range of ±30°. A further refinement is to optimise the damping of the platform at $1/\sqrt{2}$ critical, which minimises gravity meter errors, and in turn allows the operation of the gravity meter on a relatively short period platform.

The next problem to be solved is that of vertical accelerations. Any instantaneous measurement made by a gravity meter on the platform will actually be a measurement of gravity plus the vertical acceleration of the platform. No instrument can distinguish between gravity and this acceleration, therefore it is necessary to filter or average sets of measurements using a sufficiently long time constant to cancel out the vertical acceleration effects of ship motion. Furthermore, vertical accelerations make it impossible to keep the beam stationary in its null position and readings must be made with the beam in motion. In instruments designed for work at sea, the beam is very highly damped, and readings are computed from measurements of beam position, beam velocity, and beam acceleration, see figure 7/4, using the equation:

G

$$g + z'' = F \cdot S + k.B' + C \text{ milligal,}$$

where g is the force of gravity, z'' is the vertical acceleration due to motion of the platform, S is the spring tension, F is a calibration factor which converts dial readings to milligals, B' is the beam velocity (dB/dt), k is a constant which is a function of the damping coefficient, and C is the cross-coupling correction (see below). The value of z'' averages to zero over the period of the time constant. The automatic measuring system controls spring tension through the beam nuller, which operates as a slow servo-system, adjusting spring tension to

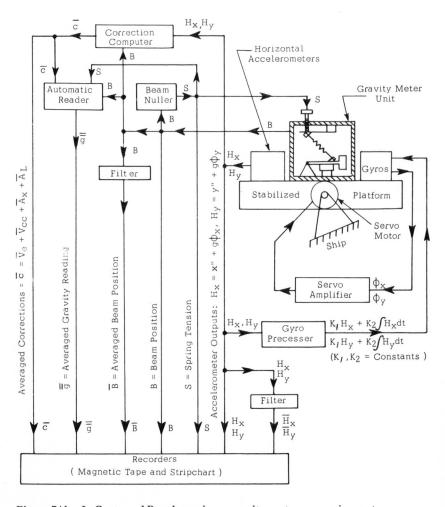

Figure 7/4   LaCoste and Romberg air-sea gravity meter measuring system.

approximately null the beam, but doing so sufficiently slowly so that sea wave accelerations are not followed.

The system so far described provides a stable platform, and a means of coping with time-varying vertical acceleration. It remains to take account of the effects on this type of gravity meter of horizontal acceleration. The two horizontal accelerometers (figure 7/4) have dual functions; that of precessing the gyros, and that of providing horizontal acceleration data to the cross-coupling computer.

Errors due to cross-coupling between horizontal and vertical accelerations are inherent in the operation at sea of a gravity meter of this type with a beam hinged about a horizontal axis. (Other meters described below do not suffer from inherent cross-coupling.) Cross-coupling errors result from the fact that in a ship, horizontal and vertical accelerations due to wave motion have the same period. Consequently, as the beam is driven by vertical acceleration, it will oscillate with the same period as that of the variation of horizontal acceleration along the axis about which it is hinged. Under such conditions, the effect of periodically varying horizontal acceleration does not average to zero and a correction must be applied which is a function of beam position and horizontal acceleration. The correction computer (figure 7/4) applies this correction as well as compensating for imperfections in the gravity meter.

Cross-coupling corrections $\bar{c}$, spring tension S and beam velocity B are fed to the automatic reader, see figure 7/4, which computes the gravity reading g, then averages this for a selected time constant, the filtered gravity output being fed to a strip chart recorder. A number of other parameters are similarly recorded to monitor the meter's performance and for use in estimating probable accuracy. Because of the complexity of the system and the need to indulge in considerable number crunching before the set of gravity readings, as recorded on an analogue chart, can be used by the geophysicist, the gravity data is usually digitised and fed to a magnetic tape recorder, within a data logging system, ready for further computer reduction. Figure 7/5 shows a photograph of a complete shipboard system.

Another successful static gravity meter design, well suited to work at sea, is based on the use of a purely translatory sensor. This type of sensor is used in the Askania sea gravity meter Gss3, see figure 7/6.

Such a system is not subject to cross-coupling errors. A tube-shaped mass is guided by five fibre ligatures so that its motion is limited to $1°$.

Figure 7/5   LaCoste and Romberg air-sea gravity meter.
(Photo: LaCoste and Romberg.)

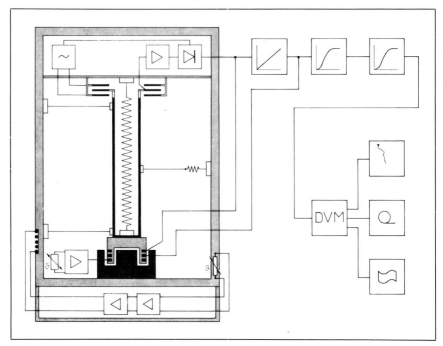

Figure 7/6   Schematic diagram of the Askania sea gravity meter Gss3.
(Courtesy: Bodenseewerk Geratetechnik GMBH.)

The mass is suspended on a mechanical spring. Changes in gravity and acceleration due to ship motion are detected by a capacitive transducer and compensation (nulling) is by an electromagnetic moving-coil force generator. Measurements are indicated directly in milligals by a calibrated digital voltmeter, and as with the LaCoste and Romberg meter, data are usually fed to both pen-recorders and to a magnetic tape data logger. For accurate results at sea, the instrument must be mounted on a very accurately stabilised platform, and, as with the LaCoste and Romberg meter, the quality of gravity data is to a large part dependent on platform performance.

In recent years, there has been much research on and development of high grade accelerometers for such applications as rocket guidance and inertial navigation systems. A shipboard gravity meter design based on use of an inertial grade accelerometer is the Bell gravity meter, see figure 7/7.

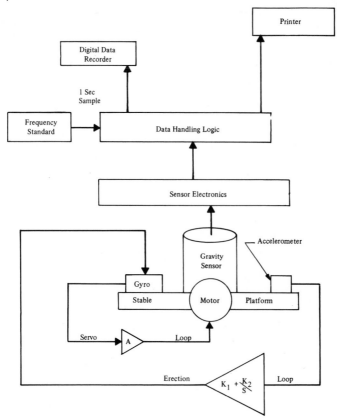

Figure 7/7 Schematic diagram of the Bell marine gravity meter. (Courtesy: Bell Aerosystems.)

The accelerometer is a single-axis pendulous force rebalance system. Because the accelerometer is itself very small, it has been possible to design this meter around a relatively small stabilised platform. Although only a few such instruments have been produced, the meters are reported to give an accuracy comparable with LaCoste and Romberg and Gss3 systems.

Vibrating string accelerometers (VSAs) are also in use as gravity sensors. In Japan, a number of TSSG (Tokyo Surface Ship Gravity Meter) instruments have been constructed using this type of sensor, as well as a few instruments which have been constructed in North America, but this type of gravity meter has not had wide commercial application in European offshore areas. The accelerometer is a dynamic rather than a static sensor; a mass is suspended on a string which is vibrated in a magnetic field, see figure 7/8.

The fundamental frequency of such a sensor is given by the equation:

$$f_0 = \frac{1}{2L} \sqrt{\frac{mg}{\rho}}$$

where L is the length of the string and ρ is its mass per unit length. Thus gravity changes, as measured by such a sensor, are indicated by change of vibration frequency: $g \propto k f^2$, where k is a constant and f the frequency of vibration. One of the major operational difficulties using such gravity sensors is that they are generally subject to fairly high frequency drifts, though with very careful construction, this can be kept to below 1 milligal per day.

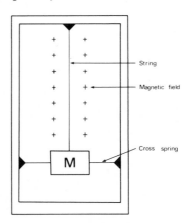

Figure 7/8 Schematic diagram of a vibrating string accelerometer (VSA).

7.5 The Eötvös correction

So far we have seen how gravity surveying depends on being able to compare an observed gravity measurement with a theoretical reference value which is a function of latitude (see section 7.2), and how equipment has been developed to observe, between stations, changes in gravitational pull, both at the seabed and on board a moving ship.

Before our final discussion of how gravity anomaly maps are prepared, one further source of error affecting gravity observations, when made on a moving ship, must be explained, and the necessary corrections described. This error is caused by the Eötvös effect which may be thought of as the vertical component of the Coriolis acceleration observed when a gravity measurement is made with an instrument which is in motion across the earth's surface. If a ship travels from west to east along the Equator at 10 knots it has an angular velocity due to the ship's speed which adds to that due to the earth's rotation, adding to the centripetal force acting on the gravity meter by approximately 75 mgals. Similarly, if another ship were travelling at 10 knots from east to west along the Equator, the centripetal force due to the earth's rotation would be reduced by approximately 75 mgals, and should these two ships pass closely to each other, gravity measured on the easterly travelling ship would appear to be approximately 150 mgals less than on the westerly travelling ship. If the ship travels in a true north-south direction, the Eötvös correction is zero. Furthermore, just as the centripetal force due to the earth's rotation reduced from a maximum at the Equator to zero at the poles, so does the Eötvös effect. Thus a correction must be applied which is a function of the ship's speed and heading, and to do this it is necessary to have an accurate measure of both the ship's course made good (the azimuth angle α) and its speed over ground, V, as well as the latitude ϕ of the gravity observation.

The Eötvös correction, E, is given by:

$$E = 7.503 \, V \cdot \cos \theta \cdot \sin \alpha + 0.004154 \, V^2.$$

For accurate gravity surveys it is in general necessary to measure the ship's speed over ground to an accuracy of about 0.1kt, and the azimuth of its course to about $5°$.

7.6 Gravity data reduction

As gravity meters measure only gravity differences, not absolute gravity, all surveys should be linked to a station at which absolute gravity is known. Guided mainly by the influence of the International Union of Geodesy and Geophysics, there has been a high degree of international cooperation in establishing a world-wide network of such gravity base stations. This network is tied to a small number of sites where absolute gravity measurements have been made using modern, high precision equipment. In Europe, various national networks are linked by observations between airports; the UK National Gravity Reference Net 1973 (NGRN73) is in such a way linked to stations throughout Europe. This UK network is made up of one hundred and thirty-five observation points, ninety-five at Ordnance Survey Fundamental Bench Marks or Flush Brackets, seven at pendulum stations, nine at stations from within an earlier gravity net, and twenty-six at airport stations in UK and Europe. Values of gravity at each of these sites is quoted as an absolute value to 0.001 mgal with a standard error of approximately 0.04 mgal. Other national nets are similarly defined.

For control of surveys in the sea areas around the UK, the procedure adopted is to use a land gravity meter to measure the gravity difference between an NGRN73 station and a site in the port which is to be used as a base for the survey; see figure 7/9. The marine gravity meter, sea bottom or shipboard, is then observed at (or as near to as possible if on a ship) the harbour station, and all other gravity observations are quoted as gravity differences to this site. It is usual, at each port call, to check the gravity meter reading at this same site so as to be able to make any necessary allowance for instrumental drift. Gravity at any station within a survey, g_S, is thus given by the equation:

$$g_S = g_B + \Delta g_{B-H} + \Delta g_{H-S}$$

where g_B is gravity at an NGRN73 station, B; Δg_{B-H} is the gravity difference between the NGRN73 station and the harbour station, H; and Δg_{H-S} is the gravity difference measured by the sea gravity meter between the harbour and a marine station, S.

Two types of gravity map are used by geophysicists; maps of free-air anomalies and maps of Bouguer anomalies. Free-air anomaly maps are prepared from gravity data which have been corrected for latitude and

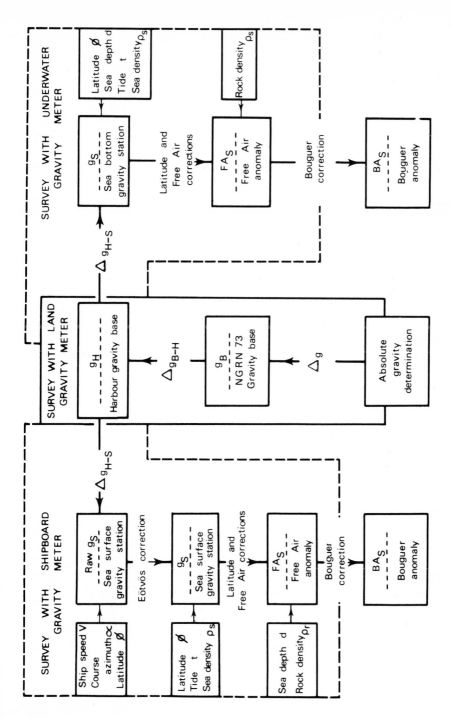

Figure 7/9 Gravity data reduction flow diagram.

elevation effects as described in section 7.2. Bouguer anomaly maps are maps prepared from data which have been additionally corrected for the gravitational attraction of the rock layer between datum and the plain of measurement, or for the gravitational attraction of the layer of sea water.

To obtain the free-air anomaly value, FA, at a station, S, the data reduction procedure is given by the equation:

$$FA_S = (g_S - g\gamma + 0.3086h) \text{ milligal}$$

where $g\gamma$ is the theoretical gravity value at latitude γ (see p. 134) and h is the height in metres of the observation above datum (usually sea-level) and 0.3086 is the elevation correction factor. This correction factor represents the free-space gravity gradient at the earth's surface in milligals per metre, due to change of distance from the earth's centre. For measurements made on the sea bottom, below a sea-level datum, the free-air correction is always of negative sign. For measurements made on a surface ship the correction is only applied if a large sea-tidal variation occurs, and can be of positive or negative sign. Continental rocks have an average density of approximately 2.67gm/cc, whereas sea-water has a density of approximately 1.03gm/cc and air may be regarded as having negligible density. The gravity field is thus affected to an appreciable degree by variation of water depth either below or above the point of observation. To compensate for this effect, Bouguer anomalies may be calculated whereby the effect is calculated of replacing the sea layer (of density ρ_s) with a rock layer (of density ρ_r).

To calculate Bouguer anomalies at sea, it is best to separate the two effects of water depth d (in metres), measured at the time of observation, and the tidal correction t (in metres), which is the height of the marine tide above sea-level datum at the time of observation. Thus for a sea bottom gravity measurement, the Bouguer anomaly, BA, at a station, S, is given by:

$$BA_S = g_S - g\gamma + d[0.04191 (\rho_r + \rho_s)-0.3086]$$
$$+ t[0.3086-0.04191\rho_r] \text{ milligal,}$$

and for a measurement made at the sea-surface by:

$$BA_S = g_S - g\gamma + d[0.04191 (\rho_r - \rho_s)] + t[0.3086-0.04191 \rho_r] \text{ milligal.}$$

For work on the continental shelf it is usual to produce contour maps of Bouguer anomalies compiled from spot gravity values which correspond either to sea bottom station determinations, or, five-minute values computed from surface ship profiles. For best results using surface ships, a line spacing of about 10km is usually regarded as sufficient to give good regional coverage, and for work over the continental margins and in the oceans a wider line spacing may be adopted.

7.7 Applications of gravity surveys

Marine gravity surveys have been widely used by government institutes, universities and other research groups to study the major structural features of the world's continental shelves and margins. As a reconnaissance method, gravity surveys give valuable data at relatively low cost on the location and sizes of sedimentary basins within basement areas. Structure of the basement itself may be studied, in particular the occurrences of high and low density igneous rocks. When used with magnetic data, an even more detailed picture can be drawn of fault patterns, and the generalised shape of structural units. Commercially, the method has not, until the last few years, been widely used in exploration for oil and gas in continental shelf areas. But with the development of major offshore oil and gas fields in areas like the North Sea the value of gravity data is becoming increasingly apparent as an aid to seismic interpretation, and it is now quite common for exploration companies to offer non-exclusive surveys which include gravity and magnetic data, as well as conventional seismic data.

The value of the method as a reconnaissance exploration tool is exemplified by the fact that many of the sedimentary basins around the UK, now the subject of extensive seismic exploration for hydrocarbons, were first discovered by non-commercial gravity surveys; these include the Moray Firth Basin in the North Sea, the West Shetlands Basin, basins in the Irish Sea, Cardigan Bay, the English Channel and the Celtic Sea.

Other applications of the gravity survey method, outside the scope of this book, include military applications associated with ballistics and navigation at sea; the requirements for gravity data in the adjustment of geodetic levelling networks; and the requirement for gravity data in the prediction of satellite orbits. Surveys undertaken primarily for military and geodetic purposes have, on a world-wide basis, provided a major contribution to knowledge of the earth's gravity field since many of these surveys have been either published or made available to research workers for interpretation and publication.

7.8 Geological interpretation of gravity data

All gravity interpretations are inherently ambiguous, and depend, in the first instance, on derivation of a geologically plausible hypothetical model involving bodies of rock of different density in the earth's crust: then the gravity field that this model would produce is calculated and this field is compared with the observed distribution of gravity anomalies obtained as a result of a gravity survey. If the observed and calculated fields show reasonable fit, the hypothesis may be considered to be tenable, and by a series of iterations the model can be modifed and refined until an acceptably exact fit is obtained. The computations required to carry out this procedure are usually complex and the literature on gravity interpretation methods is extensive. Most of these methods depend on the computational resources of large digital computers for their efficient use. Here we shall describe only how very simple and approximate estimates may be made of the size and shape of a geological structure needed to produce a particular gravity anomaly.

Before interpretation may proceed, data are required on the rock densities of the structural units involved in the hypothetical geological model. In some cases it may be possible to collect samples from boreholes and measure the densities of rocks in the area under investigation. If not, published values of measurements made on similar rocks from other areas must be used. It is the density contrasts between units in the structure that are of interest. Approximate values of the density of some common rocks are shown in table 7.1.

From this table we can see that a granite intrusion into gneiss will constitute a mass of low density rock in a high density region, and the effect will be a negative gravity anomaly. Similarly a salt diapir, intruded into a sedimentary succession will also produce a gravity low, or a trough of sandstone faulted into a region of metamorphic basement. On the other hand, an ultrabasic intrusion, intruded into any type of country rock, is likely to produce a gravity high.

Three simple model structures can be used to simulate in a very approximate way a wide range of geological structures. These are the sphere, the horizontal cylinder and the simple step. Such structures as salt diapirs and igneous plugs can be modelled by a buried sphere; elongate sedimentary basins, graben structures and buried ridges may be modelled by a buried horizontal cylinder; and single normal faults

Table 7/1
Common rock densities

	g/cc
Oil	0.90
Fresh water	1.00
Sea water	1.03
Unconsolidated sand	1.95-2.05
Boulder clay	1.90-2.10
Porous sandstone	2.00-2.60
Rock salt	2.10-2.40
Granite	2.55-2.65
Quartzitic sandstone	2.60-2.70
Compact limestone	2.60-2.70
Gneisses	2.70-3.00
Basalt	2.70-3.10
Basic intrusive rocks	2.80-3.20
Ultrabasic intrusive rocks	2.80-3.30

and other linear discontinuities can be modelled by a simple step. For simplicity, we shall consider only an interpretation based on a two-dimensional analysis.

In section 7.1 the sizes of spheres required to produce anomalies of certain specified amplitudes were quoted using a density contrast of 0.5gm/cc. The formula used to derive these sizes was as follows:

$$\Delta g = 4/3 \ \pi \ R^3 G.\rho. \ \frac{\cos \theta}{r^2} \ \text{milligals},$$

where R is the radius of the sphere, G the gravitational constant, ρ the density contrast, with θ and r defining the position of the observation point in relation to the centre of the sphere as shown in figure 7/10(a) which shows the shape of a gravity anomaly profile across such a buried sphere. A profile across a buried, infinitely long, cylinder is of very similar shape to that across a buried sphere, though of course the anomaly pattern does not have circular symmetry on a map as would be the case with a sphere. The gravity anomaly corresponding to a profile perpendicular to the axes of the cylinder is given by:

$$\Delta g = \frac{2\pi \ G \ R^2.\rho.d.}{r^2} \ \text{milligals},$$

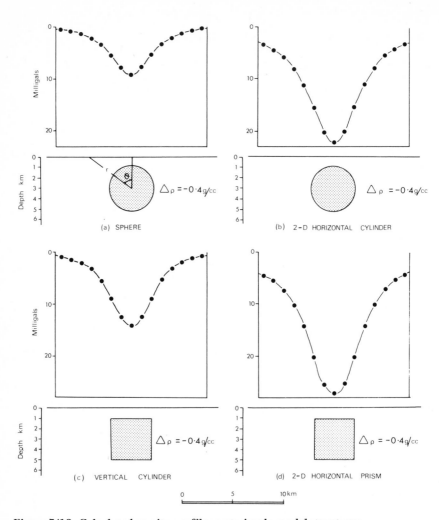

Figure 7/10 Calculated gravity profiles over simple model structures.

where d is the depth of the cylinder's axis below datum. In figure 7/10(b) a profile is shown across a horizontal cylinder of the same radius R, density contrast and depth of burial as the sphere in 7/10(a). In figures 7/10(c) and (d) the anomalies are shown across further similar models with the same density contrast; in (c) across a vertical cylinder and in (d) across an infinitely long square-sectioned prism. It should be noted that in both cases, (a) compared with (b) and (c) compared with (d), the elongate model produces an anomaly of twice the magnitude of the circular symmetrical model.

The formula from which a gravity profile across a fault or step structure can be calculated is a little more complex:

$$\Delta g = 2G\rho\,[\,x.\ln{}^{r_1}\!/r_2 + D(\pi-\theta_2)-d(\pi-\theta_1)\,]\ \text{milligal},$$

where d and D are the respective depths to top and bottom of the step. In figure 7/11 the profile across a step is shown where ρ is as above in the figure 7/10 models, and where (D-d) equals 2R. It can be seen from the above that the total gravity change across the fault is given by:

$$\Delta g = 2\pi G\rho(D-d)\ \text{milligals},$$

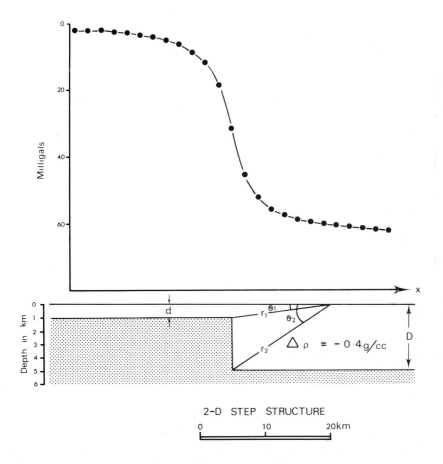

Figure 7/11 Calculated gravity profile across a model step structure.

The simple formulae given here,allow only a crude assessement of the relationship between geological structure and observed gravity anomalies. Final interpretations are usually accomplished by computer modelling.

To illustrate how the method is applied, we shall see how these formulae can be used to evaluate geological structures associated with an actual example of a gravity map of the continental shelf area in the UK. The gravity map is shown in figure 7/12. It is a Bouguer gravity map based on a survey of the Moray Firth area made by the Institute of Geological Sciences and includes both land and sea data. The marine survey was made using a LaCoste and Romberg air-sea gravity meter operated along a grid of lines of average spacing 10km. The main feature of the map is the pronounced gravity low which occupies the Firth, terminated to the north by a strong linear feature of steep gradients. A qualitative interpretation suggests a basin of low density

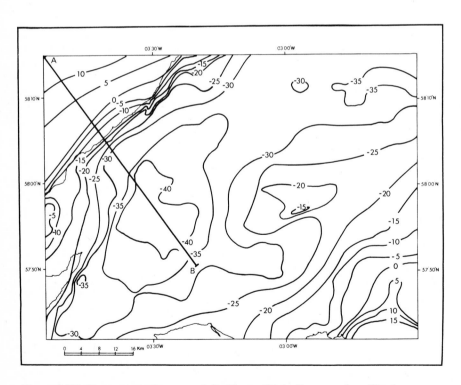

Figure 7/12 Bouguer gravity map of the Moray Firth. Contours in milligal.
(Adapted from Chesher and Bacon, 1975. 'A deep seismic survey in the Moray Firth'.)

sedimentary rocks bounded by faults to the south, the north and the west, and opening eastwards into the central North Sea trough.

As a first estimate of the size of this basin, it is possible to compare an observed profile across one of the faults features with a profile A-B computed for a simple step model.

A density contrast between basement rock and the lighter sedimentary rocks inside the basin is assumed of −0.4gm/cc, this being as high a contrast as is likely to occur thus giving a minimum estimate of the thickness of sediments in the basin. In figure 7/13 the two profiles are compared. The fit is not particularly good, suggesting that the fault feature is not exactly represented by a simple vertical step. In a published interpretation of this anomaly by Sunderland, a more complicated model has been constructed using a computer method and this is shown in figure 7/14. Although this gives a closer fit and thus more information on the shape of the basin, it can be seen that the analysis based on the simple fault model gives a comparable value for the thickness of younger sedimentary rocks enclosed within the basin.

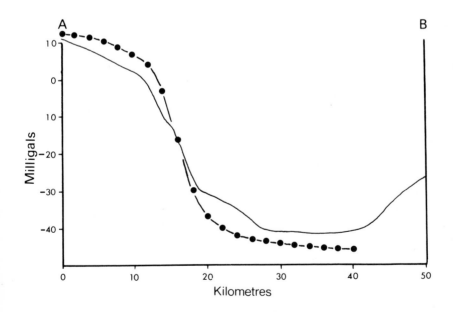

Figure 7/13 Computed (dots) and measured gravity profiles across margin of Moray Firth sedimentary basin. Model is a 4km step identical to that illustrated in figure 7/11.

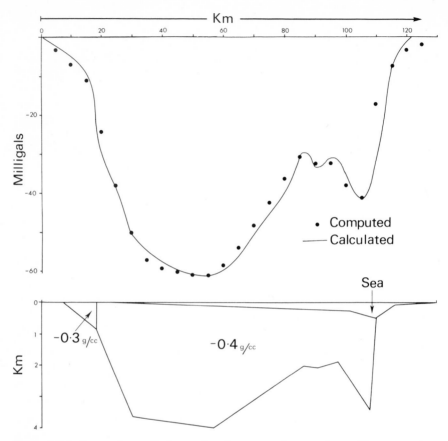

Figure 7/14 Comparison of observed and computed anomaly profiles across the Moray Firth for a computer-fitted model.
(After Sunderland, J. 1972. 'Deep sedimentary basin in the Moray Firth'. *Nature*, vol. 236, pp.24–25).

The history of economic exploration in this area is interesting in that it commenced with the discovery of the gravity low. At first, it was thought to be caused by a large granite pluton intruded into basement, but this early gravity survey was followed by aeromagnetic and a more detailed gravity survey, and interpretation in terms of a deep sedimentary basin became the more tenable. Subsequent seismic surveys and drilling have proved the existence of a deep trough of Mesozoic rocks and the area is now known to be the western extension of a major arm of the northern North Sea Basin. Furthermore, this area is now the scene of active exploration for hydrocarbons.

8 Sampling, Drilling and Visual Observations

In order to interpret geophysical traverse data adequately it is necessary to collect material from points on and below the seabed and, to be cost effective, the sample density has to be kept to a minimum compatible with the survey objectives, whether the requirement is reconnaissance or detailed coverage. Additional sampling may then be required in specific areas to determine precise variations in geological and geotechnical parameters or to conduct a prospecting evaluation. To achieve a full understanding of the geology in any given area of the continental shelf one must conduct a sequence of operations subsequent to the acquisition of magnetic, gravity, side-scan and profiling data which comprises bottom sampling, coring, drilling, visual inspection and, where relevant, geotechnical measurements.

It is beyond the scope of this review to consider hand-operated equipment, the deep drilling conducted for the oil industry, or techniques used in the measurement of geotechnical parameters, a discipline currently subject to much critical debate.

All of the activities to be undertaken demand the highest possible efficiency in application of sampling techniques and the handling of equipment, as the principal cost involved is invariably that of occupying the sample station. The mode of navigation selected has to provide both a means of occupying a pre-selected site and a precision adequate to integrate the data gathered in the overall interpretation. To this end a high degree of formality should obtain in recording information from any station. The following should be noted in addition to the sample or data description: date and time; descriptive location; ship; all primary navigation input, plus corrections where appropriate; computed latitude and longitude; water depth as measured together with corrected depth; geologist; equipment used; number of bags, bottles or cores achieved.

For many types of geological work at sea a specially designed ship is not essential; the design must be suitable to allow easy handling of equipment, the vessel must have stability and manoeuvreability and often a well-developed anchoring capability.

The use of a suitable vessel and well-organised and drilled handling procedures ensure that bad weather conditions impose the minimum delay, provide equipment with a safeguard against damage or loss and help to create conditions for sample retrieval which cause minimal disturbance. Finally, shipboard processing and storage, and transit to the onshore laboratory, should receive the degree of attention often only given to the act of collection.

8.1 Sediment sampling

A great variety of methods have been employed in taking sediment samples, ranging from tallow on the lead line, used when taking depth measurements prior to the introduction of the echo sounder, through to coring devices of increasing complexity.

In the reconnaissance surveys carried out by IGS, where sampling follows comprehensive geophysics traversing on a 5km grid, the usual pattern is to take grab and sediment cores at the intersections of the grid and in areas of transition between various bottom types. This pattern, when linked to a more detailed appraisal of type localities, provides a broad base of data from which a geological interpretation can be derived.

8.1.1 Dredges, grabs and box corers

Samples of sediments obtained by dredging suffer from several disadvantages. They are not derived from one point but from passage across a stretch of seabed, and during the haul to the surface, the contents of the dredge are subject to washing with consequent loss from the fine fraction, furthermore, little protection is afforded against loss during recovery onto the ship. Various types of dredge exist, their virtues being simplicity of construction and the large bulk of sample obtained.

Where the objective is to sample material which is distributed on the sea bottom, for example glacial erratics or manganese nodules, then the technique is appropriate and one might even equate the quantity gathered against the area traversed in a rude attempt to obtain a measure of concentration.

The design of the dredge can vary between the small, solid, ovoid cone types to large rock dredges fitted with mesh and chain link bags up to 1m deep by 1m wide. It is usual for most types to have teeth along the upper and lower frontal edges and to be connected to the towing wire by means of an A-bracket which permits only limited articulation of the dredge mouth in order that this be controlled perpendicular to the sediment during excavation. It is important to fit the larger types with a weak link at the shackle point and a reversal wire to the tail of the bag or, alternatively, a shear pin to permit full articulation of the A-bracket and overturn of the mouth in the event of severe snagging.

Grabs consist of buckets or segments which drive into the sediment and enclose and retain a sample. In general the greater the number of buckets or segments the less chance there is that total closure will be achieved due to a shell, pebble, cobble or rock fragment being held in the part open jaw. The orange peel and clamshell styles have multiple segments while the Van Veen and its derivative the Smith-McIntyre are fitted with opposed, pivoting, quarter-cylinder buckets which are usually fitted with hinged access ports on their upper surface to facilitate sample removal. The Van Veen relies on the line pull from the ship to achieve closure while in the case of the Smith-McIntyre grab this is spring assisted. The latter incorporates a weighted frame to prevent the grab falling over on contact with the sea floor and also to provide a mass against which the springs can act to push the buckets down into the sediment before closure. A widely used type, the Shipek grab recovers a slightly smaller sample in its semi-cylindrical bucket, but its simple and less hazardous cocking procedure and its virtual complete closure provide decided operational advantages. Its format of concentric half cylinders, with the inner rotated through 180° by two external helical torque springs, when activated, creates a sample enclosure protected from washing during retrieval (see figures 8/1 and 8/2). Grain size analysis performed on material from various types of grabs indicate that the Shipek is usually more competent to obtain complete samples.

All grabs become increasingly inefficient in stiff, heavily consolidated clays or alternatively as grain size increases to include pebbles, cobbles and boulders.

Perhaps the most worthwhile improvement which can be incorporated in the grab is a camera to photograph the seabed about to be sampled.

Figure 8/1   Shipek Grab fitted with prototype sea bed camera and flash units undergoing initial sea trials. The compass bottom weight is below the water surface.
(Photo: IGS.)

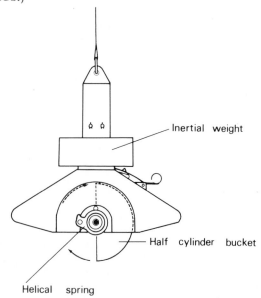

Figure 8/2   Schematic diagram of Shipek Grab.
(Courtesy: Oceaneering Engineering Corporation.)

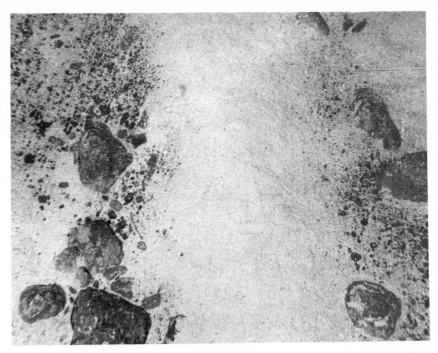

Figure 8/3   Shell sand waves traversing a gravel and pebble lag pavement. The wave is 15mm amplitude and 1m wavelength. Water depth is 77m. Location in the Sea of the Hebrides. (Photo: IGS.)

With a compass trigger weight fitted the information derived from the station photograph greatly enhances the value of the samples taken. In addition to the lithological, petrographic and palaeontologic data normally obtained evidence of bioturbation, indications of sediment movement and the relevance of the sample become apparent.

As an example of the caution which must be exercised in interpreting the product from a grab station it is perhaps worth noting the alternative samples available within the field of view when working in an area where discrete shell sand waves are migrating over a gravel lag pavement. The visual qualification of the hand sample provided by a view of the material *in situ* is therefore very worthwhile (see figure 8/3).

In order to produce larger undisturbed samples the Seckenberg box sampler used for beach sampling has been developed first to incorporate a spade to permit manual operation in shallow water and finally as a marine tool by Reineck and subsequent workers. The box sampler is more commonly used beyond the limits of the continental shelf, but is useful when minimal disturbance and a large recovery is required. However, the tool is essentially useful in clays, silts and muddy sands and not in coarse sediment types. A schematic diagram of a box sampler is shown in figure 8/4.

In routine application the Shipek grab, which produces a volume of sediment up to 3000cc, requires the smallest winch and davit capacity, normally being deployed on a 6mm diameter steel wire, while the box corer requires heavier lifting facilities and is therefore more restricted in use by available handling systems and deteriorating weather.

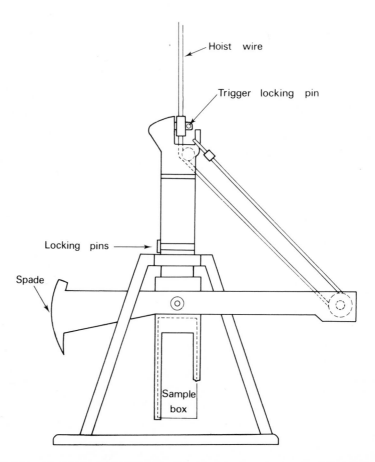

Figure 8/4   Box corer. A stylised illustration showing the principle of operation. Of this particular design, two sizes are available; one is 7ft 5in high with box dimensions of 12 x 24 x 8.5in and a weight of 1000 lb. Also, there is a half-scale model weighing 95 lb.
(Diagram based on an Oceanographic Inc. system.)

8.1.2 Gravity Corers

Lateral variation in sediment type is usually very limited when compared with vertical change and therefore it is important to determine the stratigraphy to the greatest possible depth at every station. Operations within continental shelf depths make the gravity corer the most economical coring device for this purpose when the sediment conditions are suitable. The ideal coring sequence is a fast, uninterrupted push, with adequate venting in order not to hinder movement of sediment into the barrel. No rotation should take place during penetration but, after a rest period to allow development of adhesion and friction between the sample and tube, rotation should be applied before a smooth, slow, uniform withdrawal takes place. However, in the marine situation neither a pause nor rotation are possible, as the ship cannot be held precisely on station, nor can torque be applied via the hoist line. The slow heave is attempted but is influenced by the vertical motion of the ship.

The use of a free fall winch gives controlled pay-out of the hoist line and the attainment of terminal velocity by the corer prior to contact with the seabed. In practice, by trailing the brake once the corer chassis is below the sea surface, time is saved by a virtual free fall through the water column. The drag of the line from the winch, if necessary aided by the brake to prevent the development of any snaking, is sufficient to prevent corer misalignment. Similarly, the use of fins is unnecessary as they only serve to increase the unit length to be handled aboard ship and therefore increase the required lifting height and the deck area utilised.

For launch and recovery of gravity corers, IGS has developed a trough, compatible with A-frame, davit, or boom lifting facilities, which increases the safety of the operation, allows stations to be occupied in bad weather conditions, and provides the most gentle recovery conditions in order to minimise core disturbance (see figure 8/5). This is used with half ton gravity corers fitted with stainless steel, butterfly non-return valves. It is estimated that they reach terminal velocity in free fall conditions in 20m, although their precise drag coefficient has not been measured.

The static and kinetic energy available to drive the core barrel into the seabed are derived from the weight and velocity of the coring unit. In continental shelf situations the alternative methods of driving in the core barrel are not competitive as the means of handling a suitable weight is usually easily provided. The application of piston coring as developed by Kullenberg, using a trigger weight to give a controlled amount of free fall and a static piston to improve recovery is not

Figure 8/5 An IGS gravity corer system showing the half ton chassis unit cradled in a flared launch trough with the housing line attached ready for recovery to the inboard position.
(Photo: IGS.)

considered necessary in depths less than 200m. In this, the use of a relatively stationary piston, attached to a slow-moving main haul line, gives an effective but low friction seal contact with the inside of the barrel and creates an area of low pressure above the core, thus assisting the sediment to enter the barrel.

The use of explosives, or rubber in tension, as in gun corers, or hydrostatic pressure as in vacuum corers and free corers (in which a time release jettisons the chassis weight and barrel to allow recovery of the core by means of a buoyant section) introduces unnecessary complexity. Use of the latter is appropriate from a helicopter or in deep water to remove the winch problem. The former two are limited in their attainment of velocity prior to contact with the seabed.

The performance of gravity or impact corers, driving with a single thrust, is markedly limited by the nature of the sediment. In oceanic operations it is not unusual to take cores well in excess of 20m, while on the continental shelf only rarely do fine grained sediments allow cores above 6m to be achieved. As grain size increases performance deteriorates so that muddy sand is normally the most coarse which can be penetrated and recovered, while clean sand is not penetrated. Similarly, with increasing consolidation, clay becomes less readily penetrated; pebbles and boulders encountered by the cutting shoe usually prevent further progress.

In sediments through which the core barrel will normally proceed, increasing penetration raises resistance to sample entry within the barrel, ultimately raising pressure at the cutting head and preventing entry of material. The core barrel then proceeds by pushing aside sediment ahead of it.

A suite of corers exists which uses plastic storage tube as the coring device. This includes diver operated corers, the Millport corer and various types developed for use on submersibles, all of which consist essentially of a tube fitted with a non-return valve at the top, and depend on the cohesiveness of the sediment rather than any specific catcher arrangement to hold the sediment in the barrel.

Two principal areas have to be considered in corer design to achieve an optimum compromise in operation. These are the factors influencing first, penetration and second, sample entry and recovery. For best penetrative effect the thinnest wall assembly of barrel and liner tube must be achieved. The cutter then has a minimal frontal area compatible with the required space for the catcher and liner above and is given a sharp outside taper to prevent sample disturbance and entry of excess material as well as, ideally, a positive inside clearance to facilitate ingress of material.

Various methods have been used to improve entry of core. The type of core catcher has a direct influence here, as many consist of stiff blades or flap valves which are severely hindered by the medium in which they must work. These inhibit the smooth flow of core into the barrel and often damage the structure and stratigraphy of the sample. The best compromise appears to be the type with curved, overlapping, round-ended blades manufactured in thin, spring stainless steel. These blades lie back against the liner tube during penetration and give good closure during recovery. However, in stiff clays blade reversals can occur on withdrawal.

In addition to the non-return valve, or piston arrangement noted above, various sleeves have been devised in foil or plastic to envelop the core and to move up the barrel with the sample. Used in place of the more usual liner tube these virtually eliminate wall friction which increases with increasing core length. The sleeve is stored above the core nose where its container may reduce the internal diameter of the barrel. Consequently as the core passes to the less constricted area above, grain adjustments take place with a disturbance of the original fabric. The core and sleeve have also to be extruded into a suitable storage container, unlike the rigid liner tube, which acts as a transit and archiving device.

The most widely used gravity corer type for continental shelf survey is that fitted with a rigid transparent CAB plastic liner tube which provides a good container for storage of the core once cut at the top of the sample and fitted with caps taped on at top and base. Three points should be noted. One, inspection of the core can only be successfully achieved at each end prior to capping as smearing takes place on the inner surface of the tube during core entry. Two, prior to the cutting of excess liner tube and capping at the top, cores should be stored vertically to allow adjustment to surface pressure and the settling of fines from any water standing above. Three, this plastic does not prevent water loss over long periods of storage and the core and liner tube should either be sealed in plastic sleeve or coated with wax whenever this loss would influence further analyses planned.

Whenever possible, cores should be stored and transported in a vertical position, as this minimises disturbance. When laid flat, ship motion can cause washing of grains along the length of the core. As the area of core most subject to disturbance is that adjacent to the barrel wall, and also, as plug formation occurs most readily in smaller diameter barrels, the largest diameter barrel compatible with the available handling system and chassis weight should be utilised. The IGS has for most work standardised on 2½″ OD NX barrels with a liner tube giving a 2″ diameter core but 4″ OD vibrocorer barrels giving a 3½″ diameter core have been increasingly used to obtain 20ft long cores in soft clays and silty clays, especially where detailed geological and geotechnical investigation is required.

Kögler and others have extended the principle of the box corer to gravity coring by using a very thin walled, square section barrel without a liner. Relevant comparative work on coring equipment has been undertaken by Hvorslev and by Richards, Kögler and Bouma.

8.1.3 Vibrocorers

Equipment used to obtain core samples from unconsolidated sediments is required to operate in the minimum period of time, due to the cost of the ship from which it is deployed. During operation it should be as independent of the ship as possible to reduce the station keeping requirement, which can involve multiple anchoring. Furthermore, it should have a configuration and size compatible with rapid launch and retrieval, preferably without specialised handling facilities. Minimal disturbances of the sediments from their *in situ* condition is important to retain the stratigraphy intact and to provide suitable material for engineering tests. Finally, simplicity in the equipment itself can be a considerable operational benefit. The gravity corer meets these requirements well for certain sediments types, principally soft clays, muds and muddy sands. It is also very useful for coring soft rocks exposed on the seabed; for example, in the southern part of the English continental shelf. However, as in Scottish waters, the presence of numerous glacial erratics at the seabed can complicate the situation and invalidate solid sampling by this means. Use of the method is further restricted on the UK continental shelf because much of it is floored with coarser sediment, sand and gravel, or with clay of glacial or glacio-marine origin often with a gravel or boulder content.

To sample these materials vibrocorers are used. Vibratory sampling devices have been in use in Russia since the early 1950s for on-shore soil sampling and were in marine use soon afterwards in Russia and America. Electric, pneumatic and hydraulic power sources have been tried to induce the vibration or hammering effect at the core cutting shoe by rotation of eccentric weights, oscillation of pistons, or repetitive blows. As with gravity corers these techniques have been used to obtain vertical cores either in metal tubes, from which they must be extruded, or in barrels with plastic liner, which can be removed and cut into conveniently handled lengths. In addition vibration has been utilised in conjunction with air and jet lift dredging equipments which produce disturbed samples unsuitable for either detailed sedimentological analysis or for quantitative placer or aggregate evaluation, as they sample the upper layers of sediment preferentially. This is caused by the development of a depression around the drilling head with the extraction of a volume of material in excess of the core

Figure 8/6   An IGS 20ft barrel vibrocorer being launched from a skip-hoist davit. Equally successful launch procedures have been achieved by various methods including a toboggan style launch from stern handling vessels comparable to the gravity core recovery system. (Photo: IGS.)

volume. Penetration without the disturbance associated with jet or air lift equipment is therefore sought.

Pneumatically-operated corers have, in general, not been competitive with electrically-operated equipment, mainly because they involve the handling of air lines which are very subject to current drag through their large diameter, and which are usually only armoured to combat ambient pressure to 250ft water depth.

IGS has developed a range of vibrocorers using twin geared electric motors with contra-rotating eccentric weights (see figures 8/6 and 8/7).

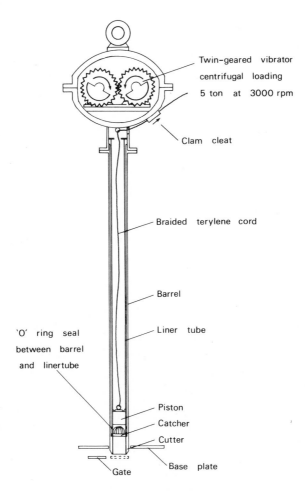

Twin–geared vibrator
centrifugal loading
5 ton at 3000 rpm

Clam cleat

Braided terylene cord

Barrel

'O' ring seal between barrel and liner tube

Liner tube

Piston
Catcher
Cutter
Base plate
Gate

Figure 8/7 An IGS piston vibrocorer. A stylised illustration of the principal elements of the drill indicating the mode of operation of the piston.

These are manufactured under licence by Aimers McLean & Co. Ltd. This contra-rotation cancels out lateral vibration and complements the vertical or longitudinal vibration. Two 415V, 3-phase motors were selected for these developments because of their immediate availability and both operate at 3000rpm. The first has a centrifugal force output between 3850 and 11,400lb depending on the setting of the eccentric weights and weighs 400lb. The second gives a centrifugal force output between 640 and 2200lb and weighs 100lb. Power is provided at 440V, to allow for the voltage drop in the umbilical, with fuse, overload and earth-leakage protection normally via a 1000ft cable. The type of frame to be used is controlled by the handling system available on the vessel and aims to provide maximum stability. Most commonly a tripod is used with one leg fitted with a canvas fin to prevent the equipment spinning and tangling the hoist and power cables. Vibrator housings have been fabricated to operate in water depths up to 1000m and frames constructed to take sample barrels of up to 10m length. Over the period of development, the frame style has remained essentially the same, except that some have been made in modular form to provide a varying capacity within the same unit. For operations from small vessels (down to 30ft length) a 5ft vibrocorer has been made using the smaller motor. This has a hinged tower with quick release guys to facilitate loading and unloading in limited deck space. It is probable that the frame could be eliminated if a simple counter-balance system was employed of the type used to support the digging head on suction dredger vessels. However, this would presumably act as a damper on the vibrator and reduce the effective penetration. In addition, it would place a more rigid requirement on the station keeping capability of the support vessel, especially now that the use of retraction equipment on the corers has minimised this.

The use of vibration rather than hammering was decided on because of the simplicity of a unit which is essentially a fixed vibrating mass incorporating the barrel assembly. In practice, it is possible to provide a hammering effect by turning the motor off and on. This creates a lower frequency, high amplitude vibration with a 27 second duration while the motor slows to a stop, which can be of value in situations where a difficult horizon is encountered in the sediment, or can be used simply to reduce external wall friction. As the barrel is not freed from the prime mover the full benefits of hammering cannot be achieved.

The barrel assembly in standard form consists of a barrel which bolts to the vibrator housing by means of a flange, a cutter which screws into the bottom of the barrel, a stainless steel core catcher with curved, round-ended, overlapping blades, a controlled piston and a C.A.B.

plastic liner tube. The latter is cut to a precise length to fit tightly between the catcher at the base and a shoulder on the vibrator housing. An 'O' ring is rolled into place between the liner tube and barrel when the unit is loaded, and this prevents the migration of sand grains into the intervening space (see figure 8/7). Prior to the use of this 'O' ring several barrels and liners, together with the enclosed samples became irrevocably locked.

A development is in hand to reduce external wall friction by fitting a water pump to the vibrocorers to provide a flow of water down between the liner and barrel, with a consequent return up the outside of the barrel via an annular set of jets at the base. This is essentially the drilling technique employed by the Beachcorer, another type of shallow coring device.

Three different barrel sizes have been used with the large vibrocorer and two with the small unit. These are a 2½" O.D. barrel giving a 2" O.D. sample and a 4" O.D. barrel giving a 3½" O.D. sample. The remaining size was an 8" O.D. barrel used without a liner tube in order to obtain large samples.

The use of 2½" O.D. barrels was abandoned for two reasons. First, it was discovered that with the small unit, plug formation was taking place in the cutting shoe and the subsequent penetration was as a probe instead of as a corer. Secondly, it seems that penetration is possible in most sediments with three exceptions. The first is a sand which has achieved or approaches its ultimate packing ratio. The second is when constituent pebbles or cobbles in the sediment exceed approximately two-thirds of the internal diameter of the barrel assembly. The third is when coring a stiff clay, a pebble is encountered which cannot be re-orientated into or out of the barrel. The second case indicates that the use of a larger barrel permits coring in a wider range of sediments. The 2½" and 4" barrel diameters were chosen because suitable liner tubing was available to fit these sizes.

The 8" diameter barrel samples successfully both through cobble horizons with cobbles in the order of 5" across, and in sand and clay. Extrusion of material from this unit is achieved with the vibrocorer lying horizontally and by applying a high-pressure water supply into the top of the barrel causing the piston to force out the sediment. Running the vibrator motor during this process enables a steady outflow of material to take place. The resulting samples are in bulk form with their detailed stratigraphy destroyed, but a recognition of the broad stratification is possible. Running the vibrator motor during

H

extraction is regarded as an emergency procedure as it invariably results in the loss of the core and an inverted catcher.

The operation of the piston, as illustrated in the sequence of operations shown in figure 8/8 helps to combat internal wall friction and aids the retention of the sample by creating an area of low pressure above the core if compaction takes place. By passing the piston line through a clam-cleat it is held at the position reached at maximum penetration during withdrawal and recovery. Pistons fitted with a valve to allow the displacement of water between their starting position above the core-catcher and the sediment surface level have been developed jointly between IGS and the UK Institute of Oceanographic Sciences.

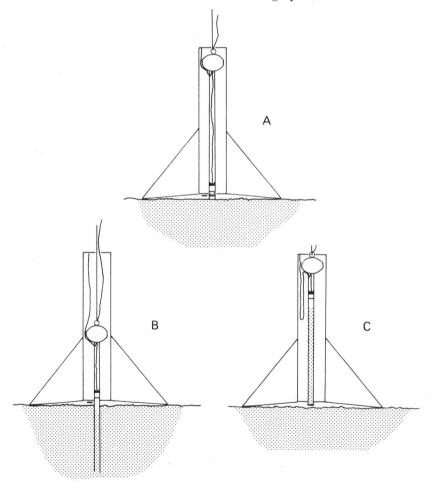

Figure 8/8   Mode of operation of a piston vibrocorer.

Alternative means of combating internal wall friction are used by some other operators. One method is to feed compressed air into the barrel until coring begins, when the air supply is turned off. There is a danger of 'air-lifting' material into the barrel with this process, ie, recovery without equivalent penetration. Another method is to control the air flow to maintain the correct ambient pressure in the barrel compatible with the penetration taking place. This use of air is essentially a shallow water option.

The performance of the equipment described above has been achieved using vibrator motors at maximum setting and with a motor and housing weight of approximately one half ton for the larger vibrocorer and 300lb for the smaller vibrocorer. An additional benefit of the large vibrocorer has been its ability to core soft bedrock material such as marls.

Complex instrumentation of such drilling tools is not necessary, but a penetration recorder can be fitted to prevent potential loss of ship time arising from uncertainty as to whether maximum reach has been achieved; without such a device, longer than necessary drilling periods need to be used. In addition this acts as a dynamic penetrometer to provide further knowledge of the engineering properties of each sample site.

Joint operations between IGS and the Bedford Institute of Oceanography, Atlantic Geoscience Centre have created instrumentation which displays and records corer attitude, current amperage, rate and extent of penetration as well as monitoring the extraction effected by an integral retraction winch. This permits use of the equipment without first having to anchor the ship which must only maintain station relative to a neutrally buoyant, braided nylon hoist line and buoyed umbilical employed in the exercise.

Vibro-hammers have been developed by Kögler at the Kiel Geological Institute with a vibrating mass attached to a barrel assembly by springs to permit direct transmission of the downward stroke while springs absorb the upward motion. Hammer corers operating at lower frequencies, with repetitive striking on an anvil at the barrel top, have been developed by other workers.

Sample handling and transportation require care to ensure that the cores do not deteriorate. For purposes of sedimentology, stratigraphy, palaeontology, geophysics and engineering, the cores should be stored

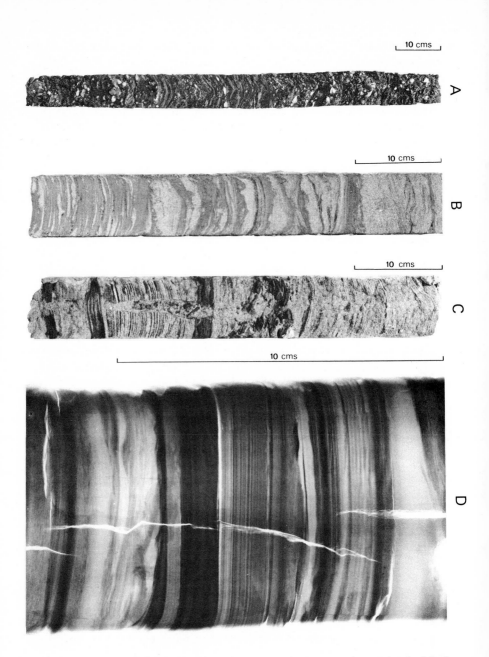

Figure 8/9   Disturbance due to the sampling process in cores obtained with the IGS vibrocorer. A and B illustrate drag structures adjacent to the barrel attributable to wall friction. C illustrates pipe structure due to fall-back of material through the catcher during recovery. D shows drag structures by an X-ray of a 1 cm slice of core.
(Photo: Dr R. Kirby, Institute of Oceanographic Sciences.)

at a temperature just above freezing point to minimise bacteriological decay and should be sealed in wax or plastic sleeving to retain their water content, as C.A.B. tubing is slightly porous. Geochemical cores require freezing immediately after sampling to prevent movement of interstitial water. For detailed study, sand and gravel cores can be impregnated with laquer to provide a relief model of internal structures, while muddy cores can be sectioned into 1 cm thick slabs and their internal structure X-rayed.

These techniques are among those which can be used to distinguish between original sediment structures and those induced by the operation of the coring device itself. The principal ones resulting from coring are:

1) marginal drag phenomena which take the form of downward curving strata adjacent to the core liner walls and which achieve a maximum in coarse sands and are minimal in clays and silts.

2) turned clasts, ie gravel pebbles, shells and well consolidated clay laminae, reorientated by contact with the cutter while entering the barrel.

3) liquefaction structures which can be further subdivided into (a) pipe structures in fine sands and interbedded sand and clay cores where incomplete catcher closure has permitted downward movement of the central portion of the core, and (b) vibration induced grain reorganisation due to fluidisation. These are structures which occur, either with a geometric relationship to the barrel, or, in lithologies where such structures are rare, or, between zones showing no such effects. Examples of small disturbances are small sand diapirs axial to the core. Homogeneous or structureless macroscopic zones also occur often extending over the upper 25 cm of core taken from material in which box coring indicates that original structure did exist.

Such effects (see figure 8/9) must be differentiated from true original structures which include slumps between undisturbed zones and biogenic homogenisation which transgress primary laminations.

It is important to establish criteria whereby the reliability of vibrocore samples may be ascertained, as they provide one of the principal means of obtaining reasonably undisturbed samples from the continental shelf. A comparison with available geophysical data and penetrometer information gives the first control. The second is obtained by a macroscopic examination of the internal structure after the opening of the cores. The third technique, applicable to sands or finer grained sediments, is magnetic fabric analysis and samples demonstrated to be reliable by this means have experienced no structural deformation, nor leaching nor rotation of grains by either mechanical means or fluidisation. However, it must be remembered that in the act of sampling, removal from the seabed will alter the stresses to which the material was subject due to the drop in hydrostatic pressure which takes place as the sample is brought to the surface.

8.1.4 *In situ* testing and drilling

It is appropriate to consider how relevant measurements of geotechnical parameters made aboard ship or in the processing laboratory are to the actual *in situ* conditions from which materials were derived. Imperfect sampling methods, change in the stress conditions following removal of the sample, hydrostatic pressure reduction during recovery, storage limitations and inadequate working conditions contribute to limit accuracy and reliability.

Offshore construction of oil industry installations commenced in 1947 in the Gulf of Mexico and has created a need for improvements in sampling and geotechnical testing techniques to meet engineering requirements which are a consequence of the increasing size, complexity and cost of structures, and the increasing water depth and worsening weather and sea conditions in which they are established. Two facts are not adequately appreciated. In onshore construction there are many occasions where the large proportion of cost escalation beyond original estimates can be attributed to inadequate site investigation. Secondly, localised site surveys need to be seen in a regional geological context in order to be interpreted adequately. The presence of a complex series of channels and buried channels in the Quaternary of the North Sea provides an example of the situation where study of a restricted area will not permit a full understanding of potential variations in ground conditions. This is critical when one considers the type of structures being installed and the static and dynamic loading which they impose on the substrate. Similarly, it was through regional studies by IGS in the North Sea that the now known wide distribution of pockmarks was discovered, these being large, shallow depression features thought to be attributable to gas emission which occur and are occurring in large numbers in certain areas, and are a significant hazard to structures, pipelines or other installations. These considerations have added considerable impetus to the basic requirement for geological sampling and several notable developments have taken place. However, because of the hostile nature of the environment and the relative instability of the platforms used for deployment, the adaption of land techniques has been limited.

Early offshore operations utilised land techniques deployed from such platforms as ships, barges or piled structures. Rotary drilling or shell and auger boring through a suspended casing extended into the seabed was supplemented by driven samples; ie equivalent to those taken by gravity coring. This technique continues to be used, but in deeper water it gives way to the use of wireline sampling as devised by McLelland Engineers Inc. in 1962. This utilised a drill string which is retracted from the bottom of the hole at the required sampling depth to allow a thin wall sample tube to be driven down into the undisturbed sediment by a hammering technique. The drilling and sampling then alternate to the required total depth of penetration required. In this type of coring operation, where the hole is liable to collapse, the use of high viscosity mud with low flow rates is critical as it stabilises the hole, minimises changes in the stress conditions within the hole, and restricts the avenue for intraformational gas escape. Typically, the driven samples are taken at 5 or 10ft intervals, but there is a need for continuous core to be taken both for a detailed appraisal of geotechnical properties and for detailed study of Quaternary stratigraphy. An attempt to core continuously through the Quaternary succession in the northern North Sea was made by IGS in 1975 from the *M.V. Sealab*, a dynamically-positioned vessel operated by Wimpey Central Laboratory Ltd. Wireline rotary drilling was employed using lightweight casing and driven samples or soil barrels, according to the nature of the sediment. A depth of 182m was reached with excellent core recovery, apart from core runs through thick layers of sand. Two features in this core may be noted; firstly, the presence of gas in certain clean sand horizons which correlate with acoustic 'bright spot' horizons give rise to an appearance akin to 'aero chocolate' in surface inspection; and secondly, the occurrence of fissuring was noted in some clay horizons particularly at depths of 58-70m. The latter causes concern with regard to the shear strength of these clays.

In the range of tools available, a gap exists between seabed equipment (eg vibrocorers), which can be operated from ships of opportunity, and drilling systems deployed from specialist ships. Increasingly, equipments of the former type are being devised to perform the dual function of sample acquisition and *in situ* testing. Similarly a range of *in situ* testing techniques have been devised for use with wireline drilling. The *in situ* tests used are mechancial (static cone penetration, vane shear and pressuremeter), or geophysical (radio-active logging, resistivity and acoustic techniques).

The Seacalf, operated by Fugro-Cesco BV is an electric cone penetrometer measuring resistance to a $10cm^2$ cone and side friction on a $150cm^2$ sleeve during continuous insertion at 2cm/s. This penetrates up to 25m depending on ground conditions, but may be limited by, for example, over-consolidation of clays. Up to 20 tons of thrust is available reacting against the equivalent weight of the assembly. In order to extend the reach of investigations the Wison penetrometer has been developed with a 90mm diameter, $5cm^2$ cone to fit within the inner diameter of drill string. This latches on to the barrel from which it derives a reaction weight of up to 3 tons. The maximum reach of 1.5m is often not attained and the interrupted measurements show a loss of precision when compared with continuous profiles.

Terresearch Ltd, have developed a remotely-operated seabed sampler weighing 10 tons which by a stepped sequence of rotary drilling, penetrometer tests and coring takes ten cores of 0.9m length to a total depth of 9m in up to 300m water depth.

McLelland Engineers Inc., in collaboration with the Norwegian Geotechnical Institute have recently developed a device known as the 'Sting Ray' which consists of a hydraulic jacking system lowered to the seabed through which a drill string is extended; an electric friction sleeve cone on a 20ft rod is locked into the lower end of the drill pipe and this is forced into the seabed by a jacking operation via the drill string. The cone rod is jacked in increments of 1-3ft until full extension is achieved and then the drill is extended when the procedure is repeated. Similarly, in co-operation with Shell Development Co., McLelland have devised a remote vane which locks into the lower end of a wireline drill string from which it derives a torque reaction; this is limited to measuring undrained shear strengths of $2.5kg/cm^2$ or less. A major virtue of penetrometers used in conjunction with wireline drill sytems is their ability to test below an obstacle or horizon which the drill has failed to penetrate.

Current developments of mechanical testing equipment include a seabed penetrometer rig by Fugro in co-operation with Shell International Petroleum Co., designed to give a minimum penetration of 15m but achieving a weight reduction by the use of suction anchors.

Fugro, together with the UK Building Research Establishment and Cambridge University are also working on a down-hole pressure meter (based on the Camkometer) to test stress-strain characteristics of critical soil layers and to determine *in situ* stress conditions.

Geophysical methods used in shallow boreholes are principally radio-active logging techniques conducted on completion of drilling prior to withdrawal of the drill string. These comprise:

1) gamma-logging, measuring the natural gamma radiation of the strata, which is usually dependent on clay content and provides a calibration of the coring and cone penetration tests,

2) neutron-logging, which is used to define moisture content, and,

3) induced gamma-logging, which is used to measure bulk density.

In situ density and porosity data are usually more reliably derived from logging techniques than from measurements on samples, but difficulties do exist in radio-activity logging due to the damping effect of observations made through the drill barrel. Also, the amount of mud filling the irregular space around the barrel can cause spurious effects.

In situ acoustic and resistivity probes have been devised to determine the velocity of the propagation and attenuation of sound, and to measure the porosity in sediments respectively. Acoustic core-logging in the laboratory appears to show a crude relationship with peak undrained shear strength.

In recent years, manned and unmanned submersibles have begun to play an important role in engineering site survey work. Along with other tasks, they have been used to deploy *in situ* testing apparatus. The main advantages gained by this procedure are precise location of the equipment and a detailed appraisal of the ground conditions by direct or remote visual inspection. It is also valuable to actually observe the tests being conducted; cone penetrometer, vane shear strength measurements, and radio-active logging have all been conducted in this manner.

8.2 Sampling outcrops and near-surface rocks

As in the case of sediment sampling the elucidation of solid geology in the marine situation ideally follows a prescribed course. A synthesis of available information, creation of detailed isopach charts and geophysical traversing, precede shallow sampling and drilling. The refinement of sampling and drilling techniques, and more especially the improved interpretations possible from geophysical data of rising quality have made it possible to reduce time on station, and to minimise the number of stations and the amount of material that it is necessary to analyse.

Three categories of equipment or effort exist for the investigation of bedrock. The first is applicable to outcrops and comprises diving and the use of small drills and submersibles. The second is applicable to the investigation of near surface sub-crop and comprises a variety of shallow penetration drills. The third (described in section 8.3) is applicable to sampling bedrock beneath a thick superficial cover and demands the use of specialised drilling vessels or large remotely operated seabed drilling machines.

Using again as an example the UK continental shelf, a wide range of rock types from hard basement to soft strata of Tertiary age exhibit the complex history of geological evolution. The varying character of bedrock exposure, coupled with the erosional effects of glaciation and the associated deposition and reworking of glacial and glacio-marine sediments, have created a situation where different techniques are required to investigate areas of outcrop in the north as compared to the south. Thus it has been possible for university workers to gather important solid rock collections from the English Channel by gravity coring and dredging whereas in the north these techniques are generally invalid, and this work has largely been achieved by use of equipment specially developed by the UK IGS.

Seabed exposures can be categorised into a number of types which, although artificial and merging into one another, indicate the variety of conditions in which sampling equipment has to operate. Consequently, as equipment has been developed from the gravity corer and rock dredge available at the outset of the IGS operations (excluding of course the existence of conventional and wire-line drilling techniques), specialisation has made individual tools appropriate for particular situations. The choice made in any given application depends on the capabilities of the ship, the weather, the current, the water depth, safety considerations and primarily, the nature of the seabed where the objective is defined. The presence of glacial erratics and gravel lag pavements, arising from the reworking of boulder clay in those areas subjected to glaciation, makes the deployment of gravity core, rock dredge or any other non-visually controlled tools inappropriate for outcrop sampling, as no certainty exists that the samples thus obtained are derived from bedrock *in situ*.

A classification of types of outcrop is worthwhile despite the arbitrary divisions:

1. Soft rock areas of subdued topography.

   (a) Largely clear of sediment other than patches of sand or mud due to current action. Here the gravity corer can be relevant if glacial erratics are absent.

   (b) With continuous thin cover of sediment which can often be penetrated by light-weight equipment.

   (c) Where an almost complete boulder moraine cover leaves only rare isolated whaleback exposures of bedrock.

2. Areas of igneous or metamorphic rocks projecting from the seabed with their steep sides fringed by boulders and the tops kept clear in shallow or exposed waters by wave action, or alternatively bearing thin deposits of sand or mud together with boulders.

3. Outcrops in shallow water formed by recent sub-aerial or marine erosion, for instance ranging from wavecut platforms in soft rock areas to gullied cliffs in hard rock areas.

4. Cliffs of hard rock, often vertical, which form the sides of over-deepened glaciated hollows. These often have sediment cover on their upper ledge, and in sheltered areas of deep water have been observed coated with mud deposits even at high angles up to 70 or 80°.

8.2.1 Bedrock sampling from conventional ships and using divers

The gravity corer system with launching trough described in the section dealing with sediment coring is suitable only in areas of bottom type 1(a). To obtain cores the chassis is fitted with heavy barrels with sharpened, mild steel, ⅜" walls which can be readily re-sharpened after use. Sample extraction is effected by driving or hammering a steel ram into the barrel from its upper end and material recovered is usually fragmented. The other tool designed primarily for sediment recovery which can obtain cores in rock is the vibrocorer. This will penetrate into soft solid rocks such as Keuper Marl and obtain short cores, even after penetrating thin sediments, and it is therefore an appropriate tool in areas of type 1(b).

To examine outcrop in areas where glacial erratics might be present scuba-divers may be used. Diving is very appropriate in depths less than 40m, especially in coastal areas where from shore or from a small boat equipped as a support platform, the mobile and versatile diver, ideally a geologist himself, can gain the required information and sample in a brief dive at a pre-selected location. Beyond this depth limit, specialist diving techniques become appropriate which are not normally cost effective for geological purposes especially as it is really necessary to use a geologist who dives, rather than a professional diver, to derive any information on the nature of the outcrop, as well as confirmation that it is *in situ* and an appropriate sample (see figure 8/10). In such circumstances, specially-designed small drilling equipment can be used. Initially, IGS utilised a small battery-operated rock drill with underwater television. This was subject to a number of limitations; ship motion was transmitted to the drill and made site location and appreciation difficult while hazarding the equipment. The drill used had a small barrel and small reach, which made final site selection difficult because of the irregularities of the rock surface and the unknown thickness of minor sediment pockets which combined to limit effective reach even further. Lateral motion could only be achieved by moving the surface ship and the whole exercise had to be conducted either at anchor or by station keeping, and the latter created the further problem that the drill might be unwittingly moved while drilling.

Experience with this equipment stimulated development along two paths. The first was an attempt to eliminate the hazard of ship motion and provide mobility, and resulted in a battery operated drill equipped with a buoyancy control system to permit the unit to hover at a controlled height above seabed, while the ship moved laterally until an

Figure 8/10 An outcrop (background) in an area of boulders with sediment occupying intervening areas. This illustrates the difficulty of locating *in situ* outcrop.
(Underwater photo at 105m depth: IGS.)

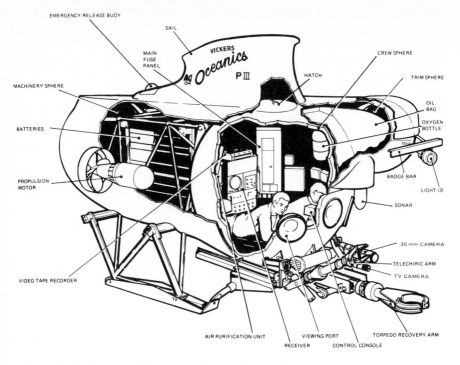

EMERGENCY RELEASE BUOY

SAIL

MAIN
FUSE
PANEL

VICKERS
Oceanics
P III

CREW SPHERE

HATCH

TRIM SPHERE

MACHINERY SPHERE

OIL
BAG

OXYGEN
BOTTLE

BATTERIES

PROPULSION
MOTOR

BADGE BAR

LIGHT (3)

SONAR

35 mm CAMERA

TELECHIRIC ARM

TV CAMERA

VIDEO TAPE RECORDER

AIR PURIFICATION UNIT

RECEIVER

VIEWING PORT

CONTROL CONSOLE

TORPEDO RECOVERY ARM

Figure 8/11 Schematic diagram of Pisces submersible.
(Courtesy: Vickers Oceanics.)

Figure 8/12 T.V. picture of sampling operation in Pisces submersible.
(Courtesy: Vickers Oceanics.)

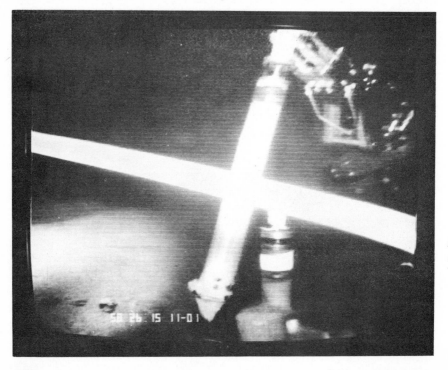

appropriate site had been located. This system was fitted with a small drill barrel with an extended effective reach and an automatic core retention arrangement, designed to take a short core after only 4cm penetration into rock. The barrel incorporated a spring loaded wedge, recessed into the barrel wall, which could be triggered to retain the core when a probe mounted 4cm behind the cutting bit contacted the rock surface. The probe rotated with the barrel before contact but was stationary thereafter when it prevented further penetration; the drill being stopped and retracted when the probe was seen to be stationary. The small drill barrel was initially developed for use on the manned submersible Pisces, operation with which also pointed to the need for a mobile visually-controlled drill, effectively independent of the support vessel, ie, an unmanned submersible system.

8.2.2 The manned submersible

Small manned submersibles are widely used in marine geology in all parts of the world. As an example of such a use we must again draw on local experience. IGS have conducted four cruises using the Vickers Oceanics Ltd submersible Pisces (see figure 8/11); three on the continental shelf and the fourth on the Rockall Plateau in conjunction with the Institute of Oceanographic Sciences. Sediment areas and rock outcrops have been examined and sampled, the objectives being the study of type localities, investigation of problem areas, examination of situations not accessible by alternative means (for example, the vertical outcrops noted in group 4 above) and geologist training.

A manned submersible presents a means whereby the geologist with little specialist training can observe and sample while conducting a traverse over the seabed and although rock climbing in a submersible has to be undertaken with due caution it provides much useful information. In addition to being an observational platform, the submersible can considerably enhance its geological role by carrying an appropriate payload. For straightforward geological survey requirements, this includes externally-mounted still and TV cameras with videotape recording facilities, the lighting arrangements being designed to give minimal back scatter both for the cameras and the occupants, as the alternative is a much reduced visibility. Sampling tools ideally include first, a multiple coring small rock drill with barrels which can be removed, either when a sample is stored aboard the vehicle and a new barrel fitted, or alternatively, in the unfortunate circumstances of a barrel jamming while drilling is in progress; second, a telechiric arm which can be used to pick up rock samples which are loose but can be seen to be *in situ*, to deploy sediment coring tubes (see figure 18/2) or to wield a bulk sediment collecting bag. This array of tools and the vehicle itself must be relatively collision-proof for operation in areas of outcrop or boulders.

Additionally, such vehicles are now being fitted with side-scanning sonar, high precision profiling and geotechnical sensors for detailed site investigation work, while their navigation ability has risen from dead-reckoning using compass reading and elapsed time, to precise location by interrogation of bottom transponder arrays. The very high cost involved in using manned submersibles necessitates a stringent planning procedure to ensure best use of the available time on site. A preparatory detailed geophysical investigation by depth-recorder, side scan sonar and sub-bottom profiler is appropriate, so that any traverse undertaken has a clearly defined objective, course and expectation. Similarly, debriefing, tape editing and report preparation should immediately follow the dive in order that the maximum information be abstracted from the exercise.

8.2.3 The remotely-operated submersible

The geological benefits of manned submersible operations are well established but the facility can only be available on a restricted time basis because of the high costs involved. In many circumstances, unmanned submersibles can be more cost-effective. An example of a remotely-operated submersible is that designed and constructed for IGS by the British Aircraft Corporation Ltd (BAC) with joint funding from the UK Department of Industry and technical assistance from the Marine Technology Support Unit of the Atomic Energy Authority, Harwell.

The object was to produce a vehicle which would be available on a continuous basis, capable of being deployed from a 'ship of opportunity', but with particular reference to use from a shallow drilling vessel as a poor weather option to be used when conditions prevented drilling. The vehicle was conceived as a means of traversing and sampling the seabed while isolated from the motion of the ship. The performance specification required an area of operation of 50m radius around a clump block in a 2 knot current in depths of up to 600m, but, in operations to date, Consub, the vehicle created, has been deployed without the block but using a buoyed umbilical. The payload fitted to the vehicle consists of two TV systems, one static for use by the driver and one mounted on a pan and tilt assembly together with a stereo still photography system and a small rock drill. Items carried within the payload can be varied and colour TV equipment has been fitted to the vehicle as an option in place of the rock drill for certain specific dives (see figure 8/13 and 8/14).

The geologist is thus provided with a shipboard simulation of the manned submersible situation, viewing the seabed and taking samples in the comfort and relative security of a surface ship operation room, where he can more readily discuss with colleagues problems encountered in real time, while still having the facility to review them later by means of video tape replay.

The Consub vehicle consists of an aluminium space frame of compartmentalised hollow tubing, and has positive buoyancy of approximately 50lb. Its overall size is 2.7m by 1.8m by 1.7m high and

Figure 8/13 The IGS submersible Consub showing the payload of small extendable rock drill, stereo camera system and T.V. system mounted on the pan and tilt head. The navigator's T.V. system is fitted in the vehicle frame. This precedes the fitting of remote read-out instrumentation and navigation equipment.
(Photo: British Aircraft Corporation.)

Figure 8/14 Consub unmanned submersible showing payload.

propulsion is provided by a pair of 5hp 0.5m diameter kort nozzle propellors with an identical pair providing vertical thrust. The latter are sufficiently powerful to hold the vehicle steady on the sea floor and provide a reaction force to counteract the thrust of the drill assembly.

Apart from launch and retrieval procedures (a ship to water interface problem common with all equipment) the fundamental constraint on remotely operated vehicle performance is that imposed by the umbilical cable. Handling and deployment of the umbilical are obvious considerations, but it is the cable drag imposed by currents which dictate the footprint of the vehicle. The footprint is the area over which the vehicle can operate at a given depth for any specified current conditions. When deployed from a static platform, for instance an anchored ship, in still water the footprint covers a circular area around the vessel limited by the length of the umbilical and avoidance of anchor wires or other obstacles. With increasing current the footprint becomes an area distorted downstream dependent on the power of the vehicle to combat cable drag. However, in many operations, and typically in a geological role, the submersible is required to conduct a traverse with stops and deviations en route. It is therefore desirable to have the vehicle deployed from a non-anchored vessel competent to maintain station or course, even with no or low forward speed by interplay of bow thrust with main thrust, or alternatively, by using a 360° variable direction bow thrust of the White Gill type. The latter has the advantage of not requiring a propellor to turn in the vicinity of the cable. Considerable responsibility is put on the ship's bridge control while accurate and coherent display of the relevant positions of ship and vehicle, ideally relative to geographic position and known seabed features, becomes imperative. This data must be available to both ship and submersible controllers who must co-ordinate their work precisely. The mother ship in this situation also displays a footprint limitation in that it has an area of operation relative to the vehicle position beyond which it cannot go without pulling the submersible off station.

A number of umbilical configurations can be used dependent on the station holding capacity of the surface support unit, the water depth and the current. For example, a totally buoyant cable (whether constructed by use of a kevlar strain member or by using buoys positioned at intervals along the cable length) may be used; or alternatively, a negatively buoyant cable can be suspended from the sea surface to a point near the seabed (but sufficiently clear to avoid contact through swell motion) from which a buoyant section may be deployed long enough to give an adequate radius of operation for the vehicle; or yet again, an arrangement may be used whereby a clump block is sited on the seabed from which a buoyant tether provides the mobility range required. The latter gives rise to additional handling problems as the clump block, which may be disposable, has to be accommodated, handled and launched in addition to the vehicle.

Various payload and thruster layouts are possible, for example the Snurre vehicle built by the Royal Norwegian Institute for Scientific and Industrial Research uses a 3 X 120°-displaced tiltable thrust array allied to one vertical thruster. Second generation Consub vehicles presently being built by BAC use two horizontal thrusters, one vertical thruster and one transverse thruster. Design studies by the Marine Technology Support Unit indicating the type of compromise decision inherent in selecting a vehicle layout showed that an effective performance using minimum power transmission (important in reducing cable area and thereby drag) could be achieved using a pair of tiltable thrusters, but that compared with separate horizontal and vertical control the demands on driver capability would be extremely high to achieve optimum performance.

Manned submersibles suffer from the limitation of having to incorporate life support systems and high safety factors because of the presence of personnel, furthermore, dives are limited by the duration of battery power supplies. In contrast, the remotely operated vehicle, which is essentially a propelled work-package, has theoretically infinite duration controlled by servicing intervals and the ability of the surface ship to maintain power and keep station. It is however constrained to operations within the limits imposed by its umbilical tether.

Navigation and position-fixing are a major consideration in all submersible work. In addition to the precise navigation which can be obtained by transponder arrays for detailed investigations, using either type of submersible, underway navigation systems between mother ship and vehicle are required for traverse work. Examples of detailed investigations in which submersibles have been used include microtopographic examination of gravity structure sites and the investigation of pockmarks, and here it may be noted that a resolution can be obtained from geophysical tools deployed by either type of vehicle which is not possible using surface towed equipment, and that submersibles can provide an excellent precisely-located stable platform for such use. Consub has exhibited very good stability and an ability to 'fly' steadily, hands-off, at a constant elevation. As an example of long-range traversing, not only geological traversing may be cited, but also the engineering requirements associated with investigating pipeline routes and checking pipelines themselves. BAC Ltd have developed for such requirements a submersible craft acoustic navigation and track indication equipment (SCANTIE) which portrays the relative positions of ship and submersible on a colour CRT display centred either on the ship, or the submersible, or relative to terrestrial navigation input together with additional navigation data.

Alternatives to the conventional submersible are provided by two radically different approaches. The first is the small submarine being developed by Horton Maritime Explorations Ltd based on the 93ft *Auguste Picard*. This eliminates the requirement for sophisticated launch and recovery equipment and use of a complex and expensive mother ship, by being self-supporting except for a small fishing vessel used for deployment of navigational transponders and as a safety standby unit. The vessel has a planned 1000 mile surface and 100 mile subsurface range with a maximum speed of 6 knots, a crew of six, and 90 man-days life support. Diesel power is used for surface running and charging of batteries. The operating costs relative to a manned submersible will be of interest and may be comparable. The other approach which is not likely to have immediate geological survey application because it has limited ability to traverse, is the atmospheric diving suit JIM developed by D H B Construction Ltd. This provides excellent on-site ability to view and perform numerous work tasks, but as the cross-country ability is essentially by walking, its use is more appropriate to localised investigation.

8.3 Sampling bedrock through a cover of superficial sediments

8.3.1 Seabed drilling equipment

The first class of equipment to be considered are those developed for rock drilling for longer core lengths than obtainable using the small drills described above. Such equipment can be used either on outcrop or to sample bedrock through a thin veneer of surface sediments.

An IGS equipment of this type is the 1m electric drill, figure 8/15, which carries an EX barrel powered by a 4hp, 3-phase oil immersed pressure compensated motor. This is equipped with a diamond bit, reaming shell and core spring, and takes cores of 20.6mm diameter from a 38.1mm hole. The motor drives the drill when rotated in one direction, while in the other (the reversal being accomplished by switching two phases of the power supply), a hydraulic pump drives the retraction ram to withdraw the barrel. The electro-hydraulic motor and pump assembly and the barrel are carried in a carriage running on nylon rollers in a tripod frame. A TV camera is mounted beneath the fin fitted to one of the legs to prevent the unit spinning and tangling the hoist wire, power and TV cables. It views the drill bit at the top of the field of view during the initial search for a suitable drill site, as well as the top of the drill hole, in mid-frame, during drilling.

Several revolutions of the barrel on commencement of drilling free the drill carriage from a captive nut. Its descent is then checked by a restriction in the hydraulic circuit to a rate slightly faster than the maximum likely to be approached in actual rock drilling. Flushing is provided by a pump drawing water locally. The depth limitation on this drill is imposed by the length of power and television cable through which it can be operated, and the longest fitted to date is 600m. The unit still suffers from the limitation of direct connection with a surface

Figure 8/15 IGS 1m rock drill.

vessel and therefore it becomes difficult to deploy in high swell conditions. In such conditions the television picture shows the seabed disappearing and reappearing making the search phase hazardous and incoherent. Lateral motion is only achieved by moving the ship. However, with the longer barrel and stable frame configuration, this unit has achieved a much higher success rate in obtaining cores than with the previously described smaller equipment.

The Bedford Institute of Oceanography, Atlantic Geoscience Centre, Canada, have built a successful 20ft rock drill, see figure 8/16, using the same diameter core barrel as in the IGS 1m rock drill. The greater length gives the advantage that it can be deployed without visual control in areas where over-burden is thought to be less than 5m thick. Typically, it is launched using a nylon hoist line and buoyant umbilical, both sufficiently long to allow the ship to maintain position adequately by station keeping relative to those parts of the line lying on the sea surface. Instrumentation includes attitude sensors in addition to monitors of power consumption, drill advance and retraction. The drive mechanism is an elegant arrangement with a paired threaded-nut and keyed-nut driving the exterior thread and keyway respectively on the barrel to apply rotation and downthrust, or retraction, depending on which one is driven and which not.

Another IGS development in this field is that of a combined 20ft vibrocorer/rock drill, though this has not progressed beyond a prototype stage. The mode of operation is selected prior to deployment and when fitted with the rock barrel alternative water flow rates permit the drill to jet through over-burden before coring into the bedrock beneath. The combined unit, the maxi-drill, was created because of the difficulty which sometimes arises in providing handling facilities for a number of large equipments on one vessel. It is equipped with a retraction system and three self-deploying legs which fold against the main tower to form a package which can be launched, rather in the manner of a gravity corer from a trough, especially when used on stern handling vessels. The equipment has had limited use mainly because the incorporation of a retraction system into vibrocorers, as in the 1m rock drill and the BIO rock drill, together with the increasing tendency towards non-anchored operations using nylon hoist line, have created a preference for smaller individual units, rather than the more heavy and complex combined units.

If the superficial sediment layer exceeds a few metres thickness, it is generally necessary to use a drilling ship to obtain any samples from the bedrock. However, one development which has attempted to fulfil this

Figure 8/16 Bedford Institute of Oceanography 20ft rock drill. (Photo: BIO.)

requirement of sampling bedrock from beneath moderately thick sediments, and at the same time be capable of operations in water depths up to 200m, is the Maricor drill. This has been developed by Atlas Copco Ltd for Wimpey Laboratories Ltd, and it is capable of drilling 45m into the seabed in 200m of water. The Maricor drill shown in figure 8/17, is designed as a cartridge loading, wireline drill for deeper water operations without requiring excessive standards of station keeping from the parent vessel. A reduction in the time required for laying multiple anchor moorings and rigging drill string, together with freedom from a purpose built parent vessel are also design objectives. Successful coring has been accomplished in 150m of water by the unit which weighs approximately 14 tons. Station-keeping requirements of the Maricor are less than those associated with conventional drilling operations, but each 3m increment of core has to be returned to the surface via a double taut wire connection, independent of the umbilical, and this imposes a strict limitation on ship movement. In view of the high costs associated with drilling ships capable of operating in deeper waters, it is considered that this approach to deep water site investigations might yet be fruitful, especially if down-hole measurements of engineering parameters of superficial strata, as well as borehole logging measurements, can be incorporated into the operating system.

8.3.2 Drilling ships

The final stage of geological survey reconnaissance is by drilling cored holes to identify strata at sites pinpointed in the interpretation of geophysical profiling data, knowledge of which is necessary to an understanding of the structure, history of evolution and pal-aeogeography of an area. Core material obtained allows detailed studies to be made in a variety of disciplines depending on the rock type. Just as in the oil industry where exploration drilling follows geophysical surveys, in regional geological studies shallow drilling follows other geophysical and geological investigations, generally as the final stage of the operational programme. Here we shall discuss two drilling ships used in UK waters both of which have been contracted by IGS, as well as by other operators.

The *M. V. Whitethorn* operated by Wimpey Laboratories Ltd worked on a five-year contract with IGS which commenced in 1970. During that contract it operated with a capacity to drill 200m into the seabed in water depths of up to 80m, with self-layed 6-point moorings, although this water depth was exceeded in good sea and weather conditions

Figure 8/17 The Maricor drill. Above the machinery compartment is the magazine containing drill rods and core barrels. The three support legs can be adjusted to stabilise the drill on the seabed. The sampler (not shown here) lands on the upper part of the equipment and transfers the cores from the equipment to the surface.
(Courtesy: Atlas Copco and Wimpey Central Laboratories.)

where good anchorage existed. The vessel was 80m overall length, 12m beam, had a tonnage of 1500 gross, and was equipped with a hydraulic cantilevered platform carrying a drilling rig extendable over the starboard side (see figure 8/18).

Drilling on pre-determined borehole sites constituted the priority activity throughout the programme but, in deteriorating weather and sea conditions, alternative sampling and geophysical work could be undertaken which was less weather dependent. A pattern of six Danforth anchors used for drilling could normally be laid within two hours, and the time on site for drilling was usually 36-48 hours. Wireline drilling strings with 60mm and 85mm core size were used, but the majority of drilling was conducted with a conventional cased string using normal core sizes of 76mm and 113mm. Operations were scheduled to minimise the effect of weather, with work being conducted in relatively sheltered areas in the winter months, and in exposed areas in summer when better conditions might be anticipated. The months of January and February were given over to annual refits

Figure 8/18 *MV Whitehorn* drilling ship showing over-side drilling platform. (Photo: IGS.)

and the remainder of the year was spent on continuous operations. Approximately 45% of the time was spent drilling, 33% on shallow sampling and geophysical survey while the remainder of time was accounted for by bad weather and port visits. The water depth limitation restricted operations to a small proportion of the total UK shelf; for example, off the west coast of Scotland drilling locations were restricted to shallow areas of more resistant rock, and it was generally not possible to core the sediment succession of adjacent basins and troughs. During the period of the *M.V. Whitethorn*/IGS operations it became increasingly important to core material overlying bedrock both to determine the stratigraphy and also to ascertain the geotechnical properties of the Quaternary succession. To meet the requirement for drilling in deeper water. Wimpey Laboratories Ltd commissioned the *M.V. Wimpey Sealab* which is a dynamically positioned drilling ship. This vessel is equipped with a centre well, a 33m drill derrick and a heave compensator (figure 8/19). As well as being capable of drilling successfully into bedrock the ship has also recently recovered a high quality core from the Quaternary succession in the southern part of the northern North Sea while conducting an investigation of pockmark features in that area, as mentioned in section 8.1.4 above.

Figure 8/19 *MV Wimpey Sealab* drilling ship.
(Photo: Wimpey Laboratories.)

9 Developments and Requirements

9.1 The state of the art

Warman, one of the contributors to a discussion on 'What directions should be set for hydrocarbon exploration' at the thirty-fifth meeting of the European Association of Exploration Geophysicists in 1973 expressed his view to those assembled that: 'Although you compliment yourselves on the state of your art, I think in some ways geophysics is only just emerging from the Palaeolithic era'. He supported this contention by saying that exploration only provided about ten per cent of the information needed in the search for, and definition of, recoverable oil from reservoirs. If these remarks can be fairly applied to geological exploration as a whole, and to offshore exploration in particular, then where are the main deficiencies and what improvements seem likely in the foreseeable future? Is the state of the art similarly deficient when applied to the search for other resources, and to work associated with constructional and engineering developments?

Let us consider the situation in terms of a contractor offering a service to a client. The offshore exploration geologist/geophysicist being the contractor, the client being the user of the information who may himself be a geologist or geophysicist wishing to interpret information to develop new scientific ideas, or alternatively an oilman or engineer, with more practical aims in mind. In general terms, we can consider how our various clients might specify their schedule of requirements and see how far the contractor can meet these. It will be necessary to draw the distinction between that which cannot be done at all, and that which can only be done inadequately, ambiguously or at unreasonable cost.

Commencing with the geologist as surveyor and scientist, maker of maps, classifier of information, researcher into geological processes both ancient and modern; what are his main requirements?

9.2 Geological mapping

9.2.1 At the seabed

Data are required for construction of a map of seabed sediments and rocks. Techniques should be employed to determine by remote sensing the distribution of sea floor sediment types in the range mud, silt, sand, gravel, boulders (and their mixtures), as well as to identify sediments deposited in earlier environments than of the present, such as Pleistocene glacial deposits where these occur, and to identify bedrock where this crops out. Such information is required over the complete area of study, not just along widely-separated profiles, because sediment variations are subject to abrupt changes horizontally as well as vertically, and the pattern of variation can be on a small scale compared with any survey line spacing which can normally be afforded. As part of the exploration it will be necessary to obtain undisturbed samples of seabed material for grain size analysis and mineralogical examination. It should be possible to define unambiguous relationship between sample identifications and identifications of seabed materials based on geophysical or other observations. All data must be related to a topographic map, and methods should be employed which give data not only for construction of a bathymetric map, but also on small scale topographic changes of perhaps only a few centimetres amplitude, which can be related to the effects of currents and wave motion on sediment distribution patterns.

Side scan sonar and echo sounding methods go far towards meeting these requirements in terms of giving quantitative data on topography along lines of profile, and qualitative data on both topographic and reflectivity changes across tracts of country parallel to profiles, all of which can be correlated with variations in sediment type. The value of acoustic methods would be greatly enhanced if quantitative data on variations in the frequency spectra of reflected signals could be related to the composition of the reflecting sediments. Some research is in hand, but so far no commericially-available system is in common operation. The present trend of development of sonar methods is mainly concerned with obtaining better quality data using conventional systems. In very large sonar devices used in oceanic depths, such as the Geological Long Range Inclined Asdic (GLORIA) system developed by the Institute of Oceanographic Sciences (UK), long frequency-modulated pulses at 6.4kHz are used, and return signals are processed in a correlator to give a high signal-to-noise ratio and good

Figure 9/1   Institute of Oceanographic Sciences 'Geological Long Range Inclined
Asdic'—GLORIA.
(Photo: IOS.)

resolution, in terms of the low frequency used. The design
requirements for work in ocean depths demand a large transducer
array, 5m long, and an output of about 50kW radiated power, the
system being housed in a towing vehicle which incorporates a
gyro-controlled rudder to minimise vehicle yaw, as well as an electronic
control system for steering the acoustic beam. Development of the
system shown in figure 9/1 has involved construction of a large and
expensive towed vehicle, which weighs 6.6 tonnes in air. Both launching and recovery require the service of a team of swimmers. The system
has been successfully used in continental shelf depths* to a maximum
range of 13km, but for such work it was of necessity operated at less
than full power and with modified towing and other operational

---

*   (Rusby, J S M and Revie, J. 1975, 'Long-range sonar mapping of the continental shelf'.
Marine Geology, Volume 19, pp M41-M52)

arrangements. It is significant, however, that such large range sonographs are attainable and that a range of features of sea floor topography can be resolved. There appears to be a possible area of development for sonar systems which are specifically designed for operation over ranges of about 10km on the continental shelf, though hopefully involving use of a less costly and bulky towed vehicle as in the GLORIA system.

Application of signal processing techniques to higher frequency devices (than GLORIA) is a possible means of obtaining increased range without loss of resolution thereby permitting total seabed sonar coverage between survey lines of a few kilometres separation. With available high resolution equipment it is necessary to survey lines no more than a few hundred metres apart if total sonar coverage is required. Another recent development in sonar techniques depends on making analogue tape recordings of each survey profile and replaying the scanned data on to either an oscilloscope or fibre-optic system, so that corrections can be made for all factors distorting the sonograph image, as well as differences in scale between long-course and cross-course scans. Indeed the aim is to provide in the output a sonograph which is true to scale, so that an isometric picture can be constructed showing an accurately positioned tract of seabed.

Another method, still being developed, which has potential as a means of obtaining continuous profiles across the seabed giving quantitative data on variations in sediment types, is that of continuous geochemical surveying. In recent years, a towed seabed gamma spectrometer has been developed by IGS and the Atomic Energy Research Establishment, Harwell for this very purpose. Total count, potassium, 'uranium' and 'thorium' measurements are made, and results to date show identifiable correlations between geology and radioactivity. In this, a scintillation detector is mounted in a steel probe attached to the end of a 25m flexible hose of the same diameter as the probe. This 'eel' is designed for towing with the steel probe in continuous contact with seabed at speeds of 3-4 knots and in depths of up to 200m (see figure 9/2). To maintain the eel in a correct towing attitude safely and efficiently, it is necessary to adjust the length of tow cable with change of ship speed (through the water) and with water depths. The system uses a winch which can be operated remotely from the ship's scientific or survey laboratory.

In sampling work, a number of problems remain, some specific to sediment sampling, some to sampling of both rocks and sediments, and all having a bearing on the accuracy and efficiency with which one is able to map shelf geology. The main cost involved in geological

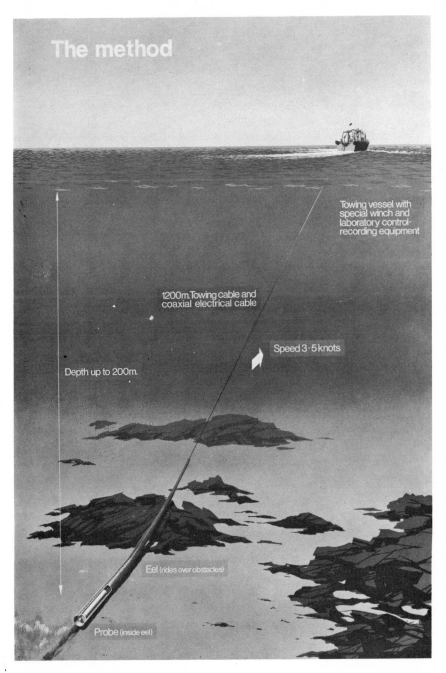

The method

Towing vessel with special winch and laboratory control-recording equipment

1200m.Towing cable and coaxial electrical cable

Speed 3-5 knots

Depth up to 200m.

Eel (rides over obstacles)

Probe (inside eel)

Figure 9/2   IGS geochemical eel.
(Photo: IGS.)

sampling is that of ship, navigation systems and personnel support. It is therefore essential that relatively low-cost sampling gear should work reliably and in as wide a range of sea conditions as possible. Development of equipment and deck-handling systems aim to provide this reliability and efficiency.

Though a range of excellent tools exist for sampling sediments, all, to a degree depending on sediment type, produce some disturbance of sample from its constitution on the seabed to that which prevails when eventually it arrives in a laboratory for testing, analysis, identification and classification. Development of equipment and techniques aim, firstly, to minimise this disturbance, and secondly, to allow discrimination between original structure and structures which can be identified as attributable to the sampling technique. Some disturbance features are not directly related to sampling technique, but to the effects of removing the sample from the relatively high hydrostatic pressure of the seabed to the lower pressure of normal atmosphere. Other causes of sample modification, ie deterioration, between seabed and laboratory include biological action and water loss. The principal need is for the development of equipment which can make a range of *in situ* measurements of as many parameters as possible on the seabed in conditions where the sediments are truely undisturbed. Such measurements could then be used to calibrate the standard laboratory analyses and tests which are likely to be less costly than tests made at the seabed.

As well as the difficulties of sample disturbance, there still remain severe deficiencies in sampling technology when applied to certain types of material. Loose sands and other coarse-grained material presents the worst problem; these can be sampled fairly efficiently at seabed using for example a Shipek grab, but coring such unconsolidated material is still a problematic task. When such problems exist, it is useful to apply geophysical borehole logging techniques which can give good control data in boreholes where core recovery is poor, as well as giving valuable data for borehole to borehole correlation purposes.

In recent years, drilling operation have become significantly more efficient due to the development of dynamic positioning systems which can hold ships accurately on a pre-determined location without the use of anchors. The benefits of the use of such systems are two-fold; firstly, it is possible to operate in a wider range of water depths without need of an anchoring tender vessel; secondly, it is possible to undertake operations in short duration weather-window conditions. The time involved in laying a spread of anchors, and the hazards involved in encountering severe weather conditions whilst anchored, between them, exert considerable restraints on sampling operations with anchored vessels.

One final note on sediment sampling; most techniques in current use exert a bias towards recovery of particular types of sediments, for example, the Shipek grab gives good recovery of loose sand at seabed, but rotary drills give generally poor recovery. It is important therefore to use a range of different types of equipment during any investigation, to ensure some compensation for this source of bias and distortion of data.

Despite the remaining problems and difficulties, the geologist has at his disposal a powerful array of tools for investigating the geology of the seabed. Yet, in many areas of the continental shelf, particularly close to coasts and in estuaries where the effects of tides and wave action are greatest, the geological pattern has such small scale variation that it remains a prohibitively costly task to map these in full detail. Furthermore, although geophysical methods give a sensitive indication of the varying nature of seabed materials, they do not give an unambiguous identification of the type of material at any particular site. It is always necessary to calibrate geophysical interpretation with sampling tests, and even when detailed sampling tests have been made, results may be unreliable because of sample disturbance.

9.2.2 The superficial sediment layer

The first requirement is to map the thickness of the layer of sediments between seabed and rockhead. A second requirement is to define the structure and variation of sediment types within this layer. To allow effective planning of sampling and drilling programmes it is necessary to be able to resolve layering close to seabed to within a few centimetres. This is particularly so in areas where thin deposits of mobile sediment or winnowed material intermittently blanket the underlying layers from which samples are required. Exploration should be able to accurately define such features as the shape and thickness of sand and gravel bodies, the internal structure of, and routes taken by, buried channels, and from these, the history of development of the superficial layer.

Geophysically, these requirements are largely met along individual profile tracks using a range of seismic profiling devices. No single device will give both adequate resolution and sufficient penetration in all circumstances, hence the wide range of devices in present day use. Trends in equipment development are concentrated at present towards the production of better seismic sources and better signal processing systems. The use of deep-towed source/receiver systems are showing significant improvement in data quality allowing penetration to many tens of metres with very high resolution of layering and structure, see figures 4/15 and 4/21 on pages 72 and 82.

Seismic profiling methods can only give data directly beneath lines of profile and regional geological mapping is often based on surveys made along a network of lines a few kilometres apart. Thus, the geological map is produced by a process of mapping detailed structures along such lines and interpolating between them. No obvious means can be seen for removing this basic uncertainty of using interpolated information, but this is a problem common to all geological mapping, on land as well as at sea.

The sediment layer can be sampled at the seabed using vibrocoring and other drilling techniques. For more detailed studies, it is necessary to obtain a completely cored section through the layer using a drilling ship. The technology of such operations is well established but subject too often to failure due to 'difficult drilling conditions'. Often, it is just those layers which present such drilling problems that are of particular interest to the geologist. There is considerable scope for improvements in present drilling techniques and the introduction of new ones, particularly because the high cost of drilling ship operations can so easily be substantially magnified by the inadequacies of methods. Alternatively, the Atlas-Copco Maricor drill, see chapter 8, which operates at the sea floor, may be a forerunner of other similar systems which can be deployed from relatively simple (low cost) ships without being dependent on good weather.

9.2.3 Bedrock

A map of bedrock geology aims to provide a synthesis of all available data appertaining to this outcropping or sub-cropping surface. Some of these data, such as dip and strike values, permit the geologist, by interpretation, to present a hypothesis on underlying deeper structure. Geophysical data and the records of deep borings allow further development of the deep structural extrapolation. The exploration requirement for production of a bedrock geological map demands the use of a range of techniques to determine all the main elements of geological structure within an area: bedding thicknesses, dips and strikes; the nature of folding with data on the trends of anticlinal and synclinal axes and their degree of plunge; the pattern of faults with data on trends, throw and hade; in sedimentary rocks, data on lithology, texture and depositional structures; in crystalline rocks, data on rock types, their deformation and jointing (and with igneous rocks on intrusive or extrusive relationships with the country rock). Even in the absence of a layer of superficial sediments, it is not possible, nor would it be practicable if techniques existed, to survey and map this geology completely to the finest possible detail over a large area. The geologist seeks an understanding of the main features and the production of a map which is generalised representation of his knowledge known to be correct within defined limits.

Geophysical methods can be used to considerable effect in surveying geological structure in bedrock where this crops out at seabed or sub-crops beneath sediments. Continuous seismic profiling using sparker, boomer, air gun or pinger sources can give excellent resolution of all the main structural elements in most geological environments. It is necessary to survey a line spacing which is smaller than the wavelength or discontinuity interval of principal structural elements such as folds, faults etc. For example, if, in a faulted area, major faults occur with an average separation of two kilometres, it will be difficult to map trends and correlate structures between lines in a grid with five kilometre spacing. If a one kilometre spacing is used there should be negligible ambiguity. In some physical and geological environments, present continuous seismic profiling methods too often perform inadequately. In shallow water, multiples can so severely interfere with primary reflected signals that resultant records are uninterpretable. In deeper water, where a thick cover of superficial sediment occurs, interference between seabed multiples and reflections from rockhead can obscure much detail. In an otherwise good environment, the occurrence of certain materials in a thick sediment layer can act as an acoustic blanket and completely obscure the underlying structures: such materials as sand and other coarse-grained sediment are most troublesome, as well as sediments which are gassified in their near surface layers. Many of these problems can be overcome by the application of digital recording and processing techniques, such as are applied in deep seismic exploration. The history of deep seismic exploration has progressed through a series of stages. In earliest days seismic records were recorded as squiggle traces, acquisition being single-channel with single-fold coverage. This stage was followed by the use of analogue magnetic tape recording and analogue processing of multi-channel data. A major step forward was marked by the development of digital recording and processing techniques, which

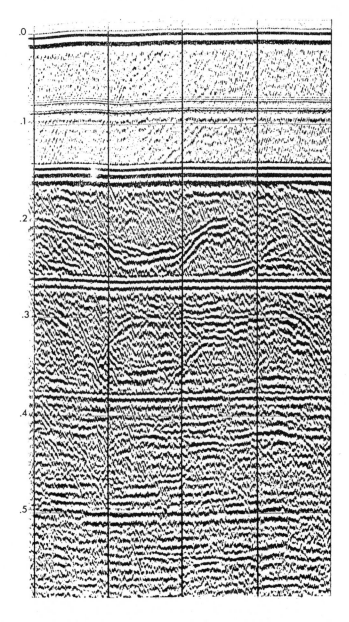

Figure 9/3 Comparison of processed and unprocessed sparker sections from the northern North Sea. Source: 3 kilojoule. The effect of multiples on the unprocessed section is such that no reliable interpretation is possible below the first seabed multiple (approximately 0.25sec two-way time).

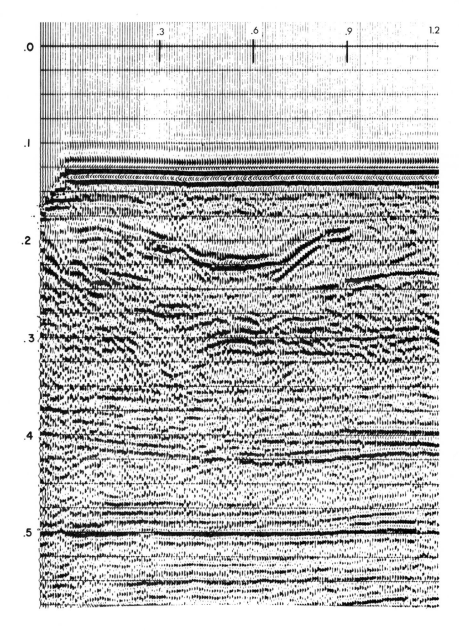

On the processed section, good reflectors can be
identified to beyond 0.6s two-way time.
(Courtesy: Aquatronics Ltd and Sun Oil Co.)

allowed multi-fold **CDP** stacking of data. So far, high resolution profiling has progressed, with but a few exceptions, only to the stage of analogue tape recording and processing. For high resolution work, high multiplicity aquisition and processing is in many respects inappropriate and even likely to reduce data quality. The trend which can be distinguished is towards application of digital recording and processing techniques using equipment specifically designed for high resolution work. Data is collected through a small number of channels at high sample rates (fractions of 1ms) and computing software is used which has been adapted from systems developed for deep seismic reflection processing. The main reason such techniques are not more widely used is that of cost; not just the capital cost of equipment, but also the high cost of processing. However, the trend towards using such techniques is likely to continue especially as advances in computer hardware technology reduce processing costs to a level where these become relatively insignificant compared with total operational costs. As an indication of the way in which a fairly simple processing package can improve sparker data, a processed and an unprocessed section are compared in figure 9/3. Of other geophysical methods magnetic and gravity have value in geological mapping, particularly magnetics when used in conjunction with seismic profiling. But, although improvements in equipment and data processing techniques might improve the use of these methods, no very significant developments are foreseen at present in the application of such methods to bedrock mapping problems.

As far as the requirement for rock samples is concerned, it is difficult to foresee a situation where the geologist will be satisfied with the amount of data he can acquire using present methods. Therefore it is necessary to exert great care in selection of sampling sites; many geological maps published today are seen to be based on very few sample identifications and even fewer borehole records. The problem is not that of not having the necessary tools but of high operational costs. The need is for tools which will reliably and quickly obtain samples, and drill holes in a wide range of sea states and weather conditions through all encounterable geological conditions at the seabed. Bedrock sampling problems are broadly similar to those of sampling in sediments; except for the need to identify outcrop and accurately locate a selected sampling device on that site. Submersibles have perhaps a special role here, in that where bedrock is exposed, they have the ability to search out such sites and sample from them; indeed the submersible provides perhaps the most reliable sampling technique in bedrock investigations at present available to the geologist other than deep drilling.

An interesting development of this role is the TOURS 430 submarine.

A design has been produced by Ingenieurkontor Lübeck for a 43m submarine, with a capability to drill on the sea floor 200m cored boreholes at diving depths of up to 500m.

9.3 Hydrocarbon exploration

This chapter commenced with a quotation by a prominent oil company representative expressing the view that exploration methods go only a little way to meet the needs of oil companies in their search for economically exploitable oil reserves. A summary of the main features of a North Sea oilfield is shown in figure 9/4. The section across the field, as well as the structural isopach map of the Brent Sand, are derived from an interpretation of seismic sections. However, this interpretation depends heavily on borehole data. It should be noted that reservoirs occur at depths of between 8000 and 10,000 feet and that the area of the field is quite small, approximately 15 x 3 miles. This field was discovered in 1971 and was the first major Jurassic discovery in the North Sea. The example is shown here to illustrate the depth of location, small size, the complex geological setting of a typical oilfield. It is perhaps not surprising, therefore, that exploration techniques might be judged inadequate. Ideally, the oilman would like to have at his disposal tools which could, in an area such as the North Sea, of approximately 200,000 square miles, unambiguously locate the relatively few oilfields, each only a few square miles in area. To do this, methods should be able to locate all potential reservoirs and be able to differentiate between structures which contain oil and gas and those almost identical structures which unfortunately do not.

Once oil or gas have been located it is then necessary to obtain for each field an accurate estimate of its size in terms of recoverable resources. In Brent these are estimated as 2,250 million barrels of oil; 3.5 trillion cu.ft. of gas. These figures could not be estimated until a late stage in the development programme, well beyond that of the original discovery. An oil company is prepared to spend large sums of money on exploration, but drilling dry exploration boreholes is not generally part of the approved programme. Each company working in a province has only so much acreage licensed to it, and the aim is to locate and successfully drill to produce oil and gas from every field that exists within its licensed area; at the same time it does not wish to waste funds and time drilling prospective-looking structures which turn out to be dry or uneconomic. In early 1976, it was estimated that the success rate for exploration drilling offshore UK was of the order one in six;

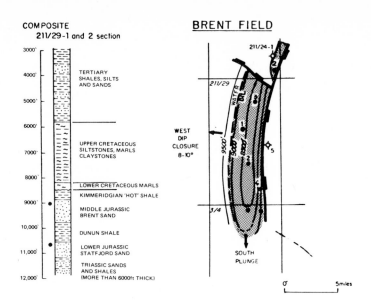

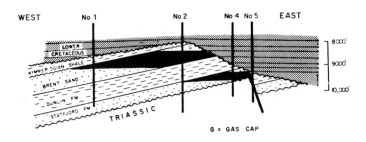

G = GAS CAP

BRENT FIELD DATA

AREA   100 - 110 sq km. (24,700 - 27,170) acres

RESERVOIR   UPPER PAY - MIDDLE JURASSIC (BRENT) SANDSTONES
                 800 ft. THICK
                 POROSITY 7-37%   PERMEABILITY UP TO 8 DARCIES
        LOWER PAY - LOWER JURASSIC - RHAETIC (STATFJORD) SANDSTONES
                 500 ft. THICK
                 POROSITY 10-26%   PERMEABILITY UP TO 5.5 DARCIES

After M Bowen (Shell) Nov. 1974 Conference
Inst Pet

Figure 9/4   Brent oil field.
(From: Pegrum, R M, Rees, G and Naylor, D, 1975. *Geology of the north-west European Shelf. Volume 2. The North Sea.* Graham Trotman Dudley Ltd, London.)

approximately 800 wells had been completed at the end of 1975 in the UK sector of the shelf, but only a proportion of these would be categorised as exploration wells.

Although all sources of geophysical and geological data are used in evaluating the hydrocarbon prospectivity of an area, it is on the basis of an interpretation of seismic reflection data that sites are eventually selected for exploration drilling, and any foreseeable marked improvement in the hydrocarbon exploration art is likely to come through improvements in the seismic method. Such improvements may come through better acquisition methods, better processing methods, or better ways of interpreting seismic data. That the method is moving towards a situation where direct detection of hydrocarbon accumulations might be possible, in some circumstances, was alluded to in chapter five when bright spot processing was mentioned. In figure 9/5, an example is shown of bright spot processing compared with conventional processing. What is sought is an association of high amplitude signals, ie with high reflective strength, with a structure which could form a hydrocarbon trap. Unfortunately, zones of high reflection strength are not all due to hydrocarbon accumulation, the indication is simply that of abnormal velocity or density change across a reflecting boundary.

Further information can be obtained using colour displays to portray on the seismic section such parameters as interval velocity and polarity. The use of colour display methods is relatively new and it is not easy to estimate just how valuable this process is likely to be in helping the interpreter to estimate the prospectivity of a structure which is a candidate for exploration drilling.

The need for higher resolution in deep seismic exploration is fully understood, but although higher sampling rates can be used, and sources designed to produce higher frequency components, the problem of attenuation of high frequencies in the rock body is a major obstacle to any great improvement in data quality through development of this approach. The alternative approach is to improve quality by a better rejection of spurious signals, by more careful elimination of all factors which distort or interfere with the process of producing a high fidelity seismic facsimile of the geological section under investigation. Developments here include the use of new methods of processing such as wave-equation processing and, under some circumstances the use of complete three-dimensional acquisition and processing.

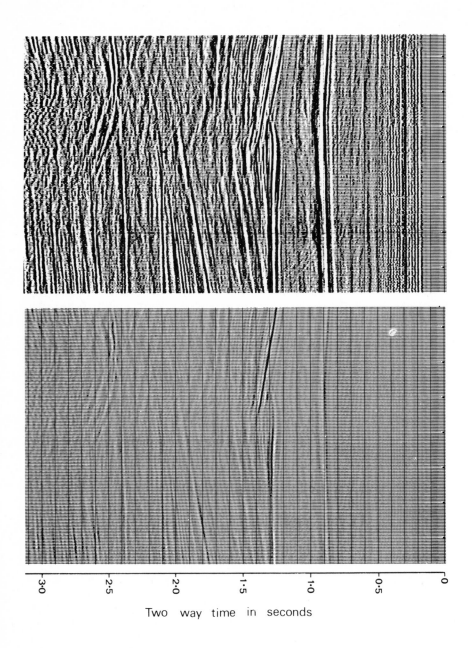

Two way time in seconds

Figure 9/5   Comparison of conventional and bright spot processing.
(Courtesy: SSL.)

Geophysical Service International (GSI) have supplied the following description of a 3-D marine seismic data acquisition and processing package operated by them:

Conventional two-dimensional seismic field geometry produces a 2-D cross-section of the response of the earth to the system used. Should any seismic reflectors under the system have a component dip in any direction other than in the plane of the cross section, then the reflectors will not be correctly represented in position on the cross-section. Equally, some information will appear on the cross-section that has arisen by reflection at points outside the plane of the cross-section. The only way to resolve this information ambiguity is to gather data in three dimensions. This implies that the source and/or detectors must be deployed in two dimensions instead of in a line. This is difficult to achieve in a marine environment and various ideas have been put forward. These include recording into several parallel streamers, or having a separate shooting boat sailing in zig-zag course along the line of a single streamer, or to cause the streamer to move obliquely across the earth's surface recording data into a band or swathe of detectors instead of into a line of detectors.

This latter approach has been adopted by GSI as a 3-D production technology and it has already been used in several surveys. The key to the technology is knowing where the air guns and the detector group centres are when each shot is made. It is also important that each swathe of depth point coverage obtained has a proper relationship to adjacent swathes and to the sub-surface problem being solved. GSI achieve this by taking advantage at the planning stage of the natural feather of the recording streamer due to tides and currents. When there is feather, the streamer moves forward relative to the water but obliquely across the surface of the earth (see figure 9/6A). Any attempt to move the streamer sideways unnaturally through the water, by paravanes for example, to induce artificial feather, is counter-productive because a large increase in ambient streamer noise is created.

The GSI streamer tracking system, keyed to the new technology, includes sensors in the streamer and on board the vessel and the outputs of these sensors permit the position of each detector group centre in relation to the boat to be computed for every shot. The geographical position of the survey vessel is computed at each shot point in post-mission processing of the navigation data. The combination of this positioning information yields, for every trace recorded in the field, an XY coordinate pair defining the mid-point between the shot and group centre, ie the depth point. The seismic traces, after appropriate normal move out and static correction have been applied, can be stacked together with much more stringent geometrical constraints than in conventional 2-D common depth point stack. An array of traces over, say, a 6 × 8 mile patch makes possible a 3-D migration stack process where for each output sample in an output trace, inputs are gathered from all the samples on a hyperboloid of revolution. In a typical example, data might be migrated into a single trace from approximately a square mile disposed around the track location. This would involve over one thousand traces to create one trace at the migration stack output stage.

Taking into account the other redundancy inherent in this multi-fold technology, it is not unusual to find 150,000 data samples being aggregated to produce one data sample when the whole processing sequence is considered.

The potential benefits of the sub-surface information extracted are clear. Defractions are correctly restored to their points of origin and complex dipping and faulted structures are accurately stacked in both a common depth point and migration sense. It is found that the better stacking geometry enhances the overall high frequency response of the system, and hence increases the potential resolving power of the seismic method.

The total cost of a 3-D marine survey is likely to be higher than the most expensive state-of-the-art 2-D survey. This is due to the close line spacing necessary to get sufficiently frequent space sampling in the cross-line direction for the 3-D (see figure 9/6; B and C), plus the vastly increased amount of number crunching required of the computer. Nevertheless, for the extra sub-surface information thereby made available, this 3-D technology is likely to become more universal in the future. Figure 9/6 shows how a programme would be planned in terms of shot line layout to allow subsequent 3-D processing.

The seismic method is still undergoing rapid development, and it would be very difficult to foresee how far it can progress before reaching a stage where new innovations, like fashionable clothes, are different to those of last season but of no greater utility. It is hoped that before this stage is reached, geophysics will be seen by the oil industry as a fully emerged post-Palaeolithic art.

9.4 Seabed engineering

It is difficult to specify a general schedule of requirements for seabed engineering purposes, for the reason that engineering projects vary so widely in scale and scope. A survey for a pipeline route will have quite different aims to that of a site survey for emplacement of a large production platform. Perhaps the first requirement of the engineer, concerned with the mechanics of rocks and soils at the seabed, is that he should have access to a good regional geological survey of the type specified in section two of this chapter. This allows him to evaluate such geological features as might cause site stability problems, create hazards, or substantially affect the design constraints to which he must work. Within a well-defined geological framework, the engineer is able to more economically define the specific requirements of a detailed and specialised site investigation.

Given that good geological and geophysical data are available, a projected engineering activity in an area then requires an interpretation of this data (with incorporation of any additional quantitative geotechnical data derived from tests at the seabed or on samples) into the form of an engineering geology map. Such a map is not generally quantitative but is a pictorial representation of all geological and geophysical and geotechnical information which has relevance to engineering requirements. Depending on presentation, one possible shortcoming of such a map is that rocks of markedly different

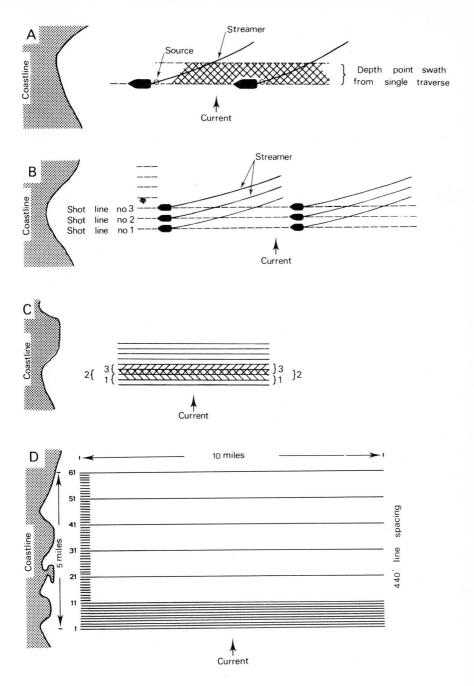

Figure 9/6   3-D seismic survey. A: single traverse showing cable drift and depth point swath. B: overlapping swaths. C: idealised overlapping depth point coverage (½ overlap). D: possible programme layout.

engineering properties can be classed within the same geological unit. As engineering activities become more widespread in continental shelf areas, the requirement can be seen for a more quantitative regional approach to engineering mapping. The need is for data on regional variations of specific geotechnical properties of seabed material. These properties can be either index parameters or design parameters; index parameters being those used to classify seabed materials into different types; design parameters being those of direct application to foundation and other engineering design work. Determination of these parameters on a close grid of stations over a large area is however, a time-consuming and costly enterprise. Apart from the cost of obtaining samples, considerable laboratory work is likely to be necessary. For completeness, a number of tests should be made on each of, say, a few hundred cores, at a number of levels beneath the seabed. These tests might involve measurements of velocity of sound, undrained shear strength, moisture content, liquid and plastic limits, natural wet density, dry density, saturated density, degree of saturation, void ratio, porosity, specific gravity and particle size distribution (some of these parameters are directly related). Contour maps can then be prepared for regional variation of any single parameter for any particular depth beneath the seabed. Some recent research has shown that ultrasonic scanning of sediment samples has particular value in this type of study and that there is a possible correlation between, for example, seismic velocity and undrained shear strength. Although detailed analysis of sample data is an essential part of such studies, it seems likely that geophysical techniques will be developed for profiling between sample sites to correlate with the results of laboratory geotechnical tests, thus permitting preparation of more detailed and more reliable geotechnical maps.

List of suggested reading

1. Belderson, R H, Kenyon, N H, Stride, A H & Stubbs, A R. 1972.
   Sonographs of the sea floor. A picture atlas. Elsevier Publishing
   Co., Amsterdam.
   A picture atlas of various types of sonographs with geological
   interpretations.

2. Bouma, A H. 1969.
   Methods for the study of sedimentary structures. John Wiley &
   Sons, New York.
   A textbook which gives information on sampling techniques and
   incorporates a comprehensive bibliography.

3. Chesterman, W D. 1974.
   'The ocean floor'. *Contemp. Phys.,* Vol. 15, No. 6, 501-516.
   A paper on the application of sonar techniques to mapping seabed
   geology.

4. Grant, F S, & West, G F. 1965.
   Interpretation theory in applied geophysics. McGraw-Hill Book
   Company, New York.
   A mathematical text on interpretation methods used in applied
   geophysics.

5. Hill, M N. 1963.
   The sea. Interscience Publishers, John Wiley & Sons, New York.
   A group of papers on a wide variety of topics in the fields of
   geophysical exploration, and sedimentation processes. The papers
   are predominantly concerned with exploration of the oceans
   rather than shelf seas.

6. Hvorslev, M J. 1949.
   *Subsurface exploration and sampling of soils for civil engineering
   purposes.* Waterways Experiment Station, Vicksburg, Mississippi.
   A classic treatise on sampling and coring methods.

7. Hydrographer of the Navy. 1965.
   Admiralty manual of hydrographic surveying. Volumes I & II.
   A very well illustrated text which covers the topics of geodesy and
   surveying methods on land and all aspects of hydrographic
   surveying.

8. Naylor, D & Mounteney, S N. 1975.
Geology of the north-west European continental shelf. Volume I.
Graham Trotman Dudley Publishers Ltd.
An easily-read text describing the geology of the western British shelf with particular reference to its potential as a hydrocarbon province.

9. Nettleton, L L. 1971.
Elementary gravity and magnetics for geologists and seismologists. No. 1. Society of Exploration Geophysicists. USA.
An elementary text with many case histories of the interpretation of gravity and magnetic surveys.

10. Parasnis, D S. 1962.
Principles of applied geophysics. Methuen & Co Ltd., London.
A concisely-written book which nevertheless explains much of the mathematical basis of applied geophysical methods within an easily understood text.

11. Pegrum, R M, Rees, G & Naylor, D. 1975.
Geology of the north-west European Continental Shelf. Volume 2. Graham Trotman Dudley Publishers Ltd., London.
The geology of the North Sea is discussed with particular reference to its development as a major hydrocarbon province. Exploration techniques are explained, including sections on drilling equipment and techniques, well-logging and production methods.

12. Richards, A F. 1967.
Marine geotechnique. University of Illinois Press, Urbana.
Though mainly concerned with the deep sea, the papers provide much useful information on the processes which control the formation of sediments on the sea floor. A section on sampling in deep water is included.

13. Sheriff, R E. 1973.
Encyclopedic dictionary of exploration geophysics. Society of Exploration Geophysicists, USA.
A well-illustrated encyclopedia covering the whole field of exploration geophysics.

14. Sly, P G. 1969.
     'Bottom sediment sampling'. *Proc. 12th Conf. Great Lakes Res.,*
     883-898, Internat. Assoc. Great Lakes Res.
     This paper describes comparative tests with a range of bottom
     sampling equipment, gravity corers, multiple corers and grab
     samplers.

15. Telford, W M, Geldart, L P, Sherrif, R E & Keys, D A. 1976.
     Applied geophysics. Cambridge University Press, Cambridge.
     A comprehensive text-book of applied geophysics providing a
     valuable work of reference for exploration geophysicists.

List of sources of technical information and illustrations

Aimers McLean & Co Ltd
Huddersfield Street
Galashiels TD1 3BB
Scotland

Fairfield Aquatronics Ltd
111 Windmill Road
Sunbury on Thames
TW16 7ES
United Kingdom

Atlas Copco GB Ltd
Swallowdale Lane
Hemel Hempstead
United Kingdom

Bedford Institute of Oceanography
Atlantic Geoscience Centre
Box 1006
Dartmouth
Nova Scotia
Canada

Bell Aerosystems Co
Buffalo
New York 14240
USA

Bodenseewerk Geratetechnik GMBH
7770 Uberlingen
Postfach 1120
Germany

Bolt Associates
205 Wilson Avenue
Norwalk
Connecticut 06854
USA

British Aircraft Corporation Ltd
Electronic and Space Systems
Filton House
Bristol
BS99 7AR
United Kingdom

British Petroleum Co Ltd
BP Research Centre
Chertsey Road
Sunbury-on-Thames
Middlesex
TW16 7LN
United Kingdom

Building Research Establishment
Building Research Station
Garston
Watford
WD2 7JR
United Kingdom

School of Physics
University of Bath
Claverton Down
Bath
United Kingdom

Compagnie Générale de Géophysique
50 Rue Fabert
75 Paris 7^0
France

Decca Survey Ltd
Kingston Road
Leatherhead
Surrey
KT22 7ND
United Kingdom

DHB Construction Ltd
Farnborough
Hampshire
United Kingdom

Euro Electronic Instruments Ltd
Shirley House
27 Camden Road
London
NW1 1YE
United Kingdom

Fenning Environmental Products Ltd (FEP)
112 Leagrave Road
Luton
Bedfordshire
LU4 8HX
United Kingdom

Fugro-Cesco BV
10 Veurse Achterweg
P O Box 63
The Netherlands

General Oceanographic Inc
Applied Marine Sciences
11578 Sorrento Valley Road
Suite 25
San Diego
California 92121
USA

Geomecanique
370 Avenue Napoleon Bonaparte
Rueil Malmaison
France

Geophysical Services International Ltd
Canterbury House
Sydenham Road
Croydon
CR9 2LS
United Kingdom

Horton Maritime Explorations Ltd
20 Brooksbank Avenue
North Vancouver
British Columbia
Canada U7J 2B8

Huntec (70) Ltd
25 Howden Road
Scarborough
Ontario
Canada

Hunting Geology & Geophysics
Elstree Way
Boreham Wood
Herts
United Kingdom

Hydro Products
Box 2528
San Diego CA. 92112
USA

Ingenieurkontor Lubeck
D 2400 Lubeck 1
Niels-Bohr-Ring 5/5a
P O Box 1690
Germany

Institute of Geological Sciences
Exhibition Road
South Kensington
London SW7 2DE
United Kingdom

Institute of Oceanographic Sciences
Brook Road
Wormley
Godalming
Surrey
United Kingdom

Institute of Oceanographic Sciences
Beadon Road
Taunton
Somerset
United Kingdom

Klein Associates Inc
Route 111
RFD 2
Salem
New Hampshire 03079
USA

LaCoste and Romberg Inc
6606 North Lamar
Austin
Texas 78752
USA

Marine Environmental Services
Harp Meadow House
62a Layer Road
Colchester
CO3 7JN
Essex
United Kingdom

McLelland Engineers Inc
6100 Hillcroft
Houston
Texas 77036
USA

North Sea Sun Oil Co
44-48 Dover Street
London W1X 4RE
United Kingdom

Nova Scotia Research Foundation
100 Fenwick Street
Box 790
Dartmouth
Nova Scotia
Canada

Polytechnic Engineering Ltd
Royal Oak Industrial Estate
Daventry
United Kingdom

S & A Geophysical
S & A House
Azalea Drive
Swanley
Kent
BR8 8JR
United Kingdom

Seiscom Delta Exploration Inc
Cumberland House
Fenian Street
Dublin 2
Ireland

Seismic Engineering Co
1133 Empire Central Dallas
Texas 75247
USA

Seismograph Service (England) Ltd
Holwood
Westerham Road
Keston
Kent
United Kingdom

Techmation
113-115 rue Lamarck
75018 Paris
France

Terresearch Ltd
Taywood Road
Northholt
Middlesex
UB5 6RL
United Kingdom

UDI Equipment Ltd
World Trade Centre
London
E1 9AA
United Kingdom

Vickers Oceanics Ltd
Barrow in Furness
Cumbria
United Kingdom

Western Geophysical
Wesgeco House
288/290 Worton Road
Isleworth
Middlesex
United Kingdom

Wimpey Laboratories Ltd
Beaconsfield Road
Hayes
Middlesex
UB4 0LS
United Kingdom

Index

Acoustic impedance   53
Acoustic methods   4
Acoustic receivers   64

Bathy metric charts   45
    chart datum   45
    tidal corrections   45
Bathymetric maps   2, 4
Bedford Institute 20ft rock drill   193
Bedrock sampling   182
Borehole velocity surveys   104
Bougier gravity anomalies   146, 148
Brent oil field   211, 212
Bright spot processing   107

Circular fixing
    see *Range-range method*
Colour display of seismic records   213
Common depth point stack   86
Continuous seismic profiling   51
    application   60, 78, 79
    interpretation of results   57, 73, 76, 79
    principles   51, 52
Core preservation   166
Corer liner tube   165

Data logging   33
Decca Main Chain
    see *Decca Navigator*
Decca Mark 12
    see *Decca Navigator*
Decibel scale   57
Deep tow seismic probing   206
Dip calculation from seismic profiles   76, 77
Direct detection of hydrocarbons   213
Diurnal magnetic variations   115
Diving surveys   4
Drag phenomena   175
Dredges   158
Drilling methods   6
Drilling ships   195
Dynamic positioning   204

Earth's magnetic field strength   116
Earth's shape   8
    ellipticity   8
    Reference ellipsoid   8

Echo sounders   32
    magnetostrictive transducers   35
    piezoelectric transducers   35
    principles of operation   32, 36
    recording systems   36
Echo sounding method   32, 33
    applications   33
    principles   33
Elevation correction factor in gravity   148
Engineering mapping   216
Engineering site surveys   80
Eötvös correction   145
Exploration costs   2

Far field signature   91, 92
Fourier analysis   124
Free air gravity anomalies   146, 148
Free-fall coring   163
Free-fall winches   163

Geochemical eel   202
Geochemical seabed surveying   202
Geological Long Range Inclined Asdic (GLORIA)   202
Geological mapping   200
Geomagnetic field   114
Geophysical logging techniques   179
    gamma logging   179
    induced gamma logging   179
    neutron logging   179
Geotechnical maps   216
Geotechnical testing   176
Grabs   158
Gravitational constant   133
Gravity anomaly maps   146
Gravity base networks   146
Gravity corers   163
Gravity data reduction   146
Gravity meters   134
    accelerometers   139, 143, 144
    Askania Gss3 gravity meter   141
    Bell gravity meter   142
    calibration   140
    cross-coupling computer   141
    cross-coupling effect   141
    dynamic meters   139
    gyro-stabilised platform   139
    LaCoste and Romberg sea bottom gravity meter   134, 136, 137
    LaCoste and Romberg shipboard gravity meter   141
    land gravity meters   134
    sea bottom gravity meters   134
    shipboard gravity meter   139
    static meters   139
    Tokyo Surface Ship gravity meter   144

vibrating string accelerometers (VSAs)   144
   zero-length spring   136
Gravity methods   5, 131
   applications   132, 149
   interpretation   150
   principles   131, 133
Gravity modelling   132, 150-153
Gun corer   165
Gyromagnetic ratio   119

Hammer corers   173
Hydrocarbon exploration   211, 213
Hydrophone Streamers   92

IGS 1m electric drill   191
IGS 20 ft vibrocorer/rock drill (Maxi-drill)   193
Impact coring   165
Induced magnetisation of rocks   124
In situ acoustic probes   179
In situ resistivity probes   179
In situ testing   176
International Geomagnetic Reference Field (IGRF)   116
International gravity formula   134
Inverse square law   114
Isochron maps   104
Isopach maps   107

JIM diving suit   191

Kögler corer   167
Kullenberg corer   163

Larmor frequency   118
Larmor's theorem   118
Liquefaction structures in cores   175

Magnetic anomalies   126
Magnetic data reduction   120
Magnetic fabric analysis   176
Magnetic methods   113
   aeromagnetic surveys   113
   applications   113
   geological interpretation   113
   marine surveys   113
   principles   113, 114
Magnetic modelling   127
Magnetic poles   114
Magnetic properties of rocks   125
Magnetic storms   115
Magnetic susceptibility of rocks   124
Map coordinate systems   8
Map datums   9

Map projections   8
     British National Grid   9
     Lambert Conical   8
     Mercator   8
     Transverse Mercator   9
     Universal Transverse Mercator   9
Maricor drill   195
Migration of seismic sections   107
Millport corer   165
Multi-channel seismic reflection surveying   85, 86
     applications   85, 108
     basic principles   86
Multiple seismic reflections   73

Newlyn datum   45
Newton's Law of Gravitation   133

Optical pumping magnetometers   117, 120
Ordnance datum   45
Outcrop sampling   180
Outcrop types   180

Penetrometers   176
Permanent magnetisation of rocks   124
Photography underwater   185
Piston covers   165, 171
Pockmarks   83, 176
Position fixing   4, 7
     accuracy   7
     principles   7
Proton magnetometers   116, 117

Radio position fixing   10
     accuracy   10
     hyperbolic method   13
     principles   10, 11
     range-range method   11
Radio position fixing systems
     Artemis   10, 23
     Cubic Autotape   23
     Decca Hi-fix   15, 16, 21
     Decca Navigator   15, 16, 17
     Decca Sea-fix   15, 21
     Decca Trisponder   23
     Hydrodist   23
     Lorac   21
     Loran-C   11, 15
     Motorola Miniranger   22
     Omega   15
     Pulse-8   15, 16
     Raydist   21
     Toran   21

Reflection seismic equipment   88
    analogue to digital computer   94
    anti-alias filter   95
    controller module   95
    digital recording   94
    hi-cut filter   95
    marine seismic sources   88
        Aquapulse   91
        Aquaseis   91
        chemical explosives   88
        Dinoseis   91
        Flexichoc   90
        Flexotir   91
        PAR air gun   88
        Sodera water gun   90
        Vaporchoc   90
    multi-channel seismic streamers   92
    oscillograph   95
    seismic camera   97
    tape transports   97
Reineck box corer   161
Rock densities   132, 148, 150, 151
Rock drills   182
Rotary drilling   177

Sample handling   173
Sampling by divers   182
Sampling methods   5
Sampling techniques   157, 163
Sand diapirs in cores   175
Satellite navigation and position fixing   24
    integration with doppler sonar   28
    integration with radio position fixing   26, 27
    principles   24
    Transit satellites   24
Seabed acoustic position fixing   28
    accuracy   28
    acoustic transponders   29
    Autranav   29
    principles   29
Seabed drilling equipment   191
Seabed engineering   216
Seabed gamma spectrometer   202
Seacalf penetrometer   178
Seckenberg box sampler   161
Sediment sampling   158, 163
Seismic picking   73, 76, 104
Seismic processing   99
    deconvolution   101
    demultiplexing   99
    normal move out   101
    time variable filtering   101
    true amplitude recovery   99
    variable area display   103

velocity analysis   101
Seismic profiling equipment
    acoustic source frequencies   55, 57
    acoustic source power output   54, 58
    analogue tape recording   70
    automatic gain control   71
    band pass filtering   71
    basic principles   52
    boomers   60, 62, 64
    graphic recorders   71
    hydrophones   64, 68, 69
    hydrophone streamers   68
    pinger probes   59, 62
    pneumatic sources   54
    recording systems   70
    sparker probes   59, 62, 64
    system designs   59, 64
    time varying gain control   71
    towing gear   71
Seismic pulse frequency   57, 60
Seismic reflection coefficient   53
Seismic refraction surveying   85, 108
    applications   111
    basic principles   109, 110
    interpretation of results   111
Seismic section analysis   103
Shear vanes   178
Shell and auger boring   177
Shipek grab   159, 162
Side scan sonar equipment   43
    basic principles   43
    display   45
    receiving amplifiers   43
    signal processing   43, 45
    transducers   43
    transmission units   43
Side scan sonar method   32, 39
    applications   39
    principles   39
    resolution   39
    scale distortion   39
    slant range distortion   39
Single point seismic reflections   73
Smith-McIntyre grab   159
Snell's Law   109
Sonic well-logging   104
Sonobuoys   111
Sonographs   39
    classification   46
    interpretation   48
Sound absorption in rocks   56, 57
Stingray sampler   178
Stratigraphic identification of seismic reflectors   103-104

Strike calculation from seismic profiles   76, 77
Submersibles   179
    Auguste Picard   191
    Consub   186, 190
    launch and retrieval   188
    manned   185, 190
    navigation   190
    Pisces   185
    Snurre   189
    umbilicals   189
    unmanned   186

Television underwater   185, 193
Three dimensional seismic surveying   213, 215
Time-distance graphs   109
Torsion balance   131
Tours 430 submarine   210
Turned clasts in cores   175
Two-range Decca   20

Van Veen grab   159
Velocity of sound
    in air   53
    in rocks   53
    in water   53, 57
Vibrocores   167, 170
Vibro-hammers   173

Wall friction reduction in corers   171, 173
Wave equation processing   213
Wison penetrometer   178
Wireline drilling   177, 178

X-raying of cores   174

Zeeman effect   120